T0134699

Studies in Systems, Decision and Control

Volume 309

Series Editor

Janusz Kacprzyk, Systems Research Institute, Polish Academy of Sciences, Warsaw, Poland

The series "Studies in Systems, Decision and Control" (SSDC) covers both new developments and advances, as well as the state of the art, in the various areas of broadly perceived systems, decision making and control–quickly, up to date and with a high quality. The intent is to cover the theory, applications, and perspectives on the state of the art and future developments relevant to systems, decision making, control, complex processes and related areas, as embedded in the fields of engineering, computer science, physics, economics, social and life sciences, as well as the paradigms and methodologies behind them. The series contains monographs, textbooks, lecture notes and edited volumes in systems, decision making and control spanning the areas of Cyber-Physical Systems, Autonomous Systems, Sensor Networks, Control Systems, Energy Systems, Automotive Systems, Biological Systems, Vehicular Networking and Connected Vehicles, Aerospace Systems, Automation, Manufacturing, Smart Grids, Nonlinear Systems, Power Systems, Robotics, Social Systems, Economic Systems and other. Of particular value to both the contributors and the readership are the short publication timeframe and the world-wide distribution and exposure which enable both a wide and rapid dissemination of research output.

** Indexing: The books of this series are submitted to ISI, SCOPUS, DBLP, Ulrichs, MathSciNet, Current Mathematical Publications, Mathematical Reviews, Zentralblatt Math: MetaPress and Springerlink.

More information about this series at http://www.springer.com/series/13304

Marcos Quiñones-Grueiro ·
Orestes Llanes-Santiago ·
Antônio José Silva Neto

Monitoring Multimode Continuous Processes

A Data-Driven Approach

Springer

Marcos Quiñones-Grueiro 🆔
Department of Automation and Computing
Universidad Tecnológica de La Habana José
Antonio Echeverría
Cujae, La Habana, Cuba

Orestes Llanes-Santiago 🆔
Department of Automation and Computing
Universidad Tecnológica de La Habana José
Antonio Echeverría
Cujae, La Habana, Cuba

Antônio José Silva Neto 🆔
Instituto Politécnico
Universidade do Estado do Rio de Janeiro
Rio de Janeiro, Brazil

ISSN 2198-4182　　　　　　　　ISSN 2198-4190　(electronic)
Studies in Systems, Decision and Control
ISBN 978-3-030-54740-0　　　　　ISBN 978-3-030-54738-7　(eBook)
https://doi.org/10.1007/978-3-030-54738-7

This Springer imprint is published by the registered company Springer Nature Switzerland AG
The registered company address is: Gewerbestrasse 11, 6330 Cham, Switzerland

To Sahai
Marcos Quiñones-Grueiro

To Xiomara, Maria Alejandra and Maria
Gabriela
Orestes Llanes-Santiago

To Gilsineida, Lucas and Luísa
Antônio José Silva Neto

Preface

The present book is intended for use in graduate courses of engineering, applied mathematics and applied computing, where tools from mathematics, data mining, multivariate statistical analysis, computational modeling, and computational intelligence are employed for solving fault diagnosis problems in multimode processes.

In this book, recent methods are presented and formalized for data-driven fault diagnosis of multimode processes. There are very little books that address the important issue of monitoring multimode systems. In some of them, the problems of dynamic and non-linear behavior as well as the Bayesian approach for fault treatment are taken into account. Nonetheless, in all cases, the processes studied only consider steady-state modes. Moreover, the data analysis tools such as clustering methods for multimode processes are not analyzed in depth.

The main objective of this book is to formalize, generalize and present in a systematic, organized and clear way the main concepts, approaches and results for the fault diagnosis of multimode continuous Processes by using data-driven methods. The main contributions of the book can then be stated as follows

- A methodology to design data-driven fault diagnosis methods inspired in the widely known Cross Industry Standard Process for Data Mining (CRISP-DM).
- A formal definition of multimode continuous processes from a data-driven point of view together with an algorithm to determine in an automatic fashion if a process can be considered as multimode based on a representative data set.
- A procedure to apply clustering methods to data acquired from multimode processes such that important features for fault diagnosis can be identified.
- A methodology to design operating mode identification and fault detection methods for multimode processes.
- A methodology to design fault cause identification methods based on data-driven classification tools.

In the book are also taken into account steady-state modes, but going further it presents fault diagnosis schemes for mixed types of modes (transitions + steady states). Thus, a more realistic approach to fault diagnosis is developed. Moreover, since multimode processes are defined formally from a data-driven perspective and an algorithm is provided, the reader can identify whether a process is multimode or not. Data analysis tools for multimode data are also presented. Therefore, the reader can select the proper tools for developing effective Fault Diagnosis schemes.

Three well-known benchmark problems in the scientific literature are used in the book to exemplify the methodology and the tools described. They are: Continuous Stirred Tank Heater (CSTH); the Hanoi Water Distribution Network (HWDN); and Tennessee Eastman Process (TEP). For the analysis of the experiments developed, several tables and graphics were constructed in order to facilitate the presentation of the results, as well as their interpretation. In addition, the Matlab code for the different methods used in this book is presented in the Appendices.

The fundamental pre-requisites to read this book are univariate and multivariate statistical methods. Some basics concepts about programming are also necessary for a better understanding of the different computational tools used in the book.

The chapters of this book are summarized as follows:

Chapter 1: **Fault Diagnosis in Industrial Systems** starts with a brief description of model-based methods, and then presents a brief review of the data-driven methods used for fault diagnosis. The fundamental concepts related to the fault diagnosis problem are presented. In the past years, the interest of the scientific community has focused towards new process monitoring problems: dynamic, time-varying and multimode behavior. Therefore, a methodology is presented to design data-driven fault diagnosis methods inspired in the widely known Cross Industry Standard Process for Data Mining (CRISP-DM). This methodology allows a systematic design of fault diagnosis methods based on a data set of the process.

Chapter 2: **Multimode Continuous Processes** presents the characteristics and definitions of multimode continuous processes. It is discussed why they require specific monitoring approaches for achieving good fault diagnosis performance. A formal mathematically definition of a multimode continuous process is presented from a data-driven point of view. An algorithm based on the definition is presented to determine if a process is multimode or not. Three different cases of study of multimode processes are presented and it is illustrated why they are multimode.

Chapter 3: **Clustering for Multimode Continuous Processes** presents the use of clustering methods for labeling the data from multimode continuous processes. Fault diagnosis represents a twofold challenge: the unlabeled data must be characterized with respect to the operating modes, and monitoring schemes must be developed by using the data once it has been labeled. The former task is challenging because of the uncertainty related to the unknown number and types of modes. Therefore, the main clustering solutions applied to multimode processes are discussed according to the types of modes. A clustering procedure is presented to allow the systematic analysis of data sets from multimode processes. It is illustrated how to use popular clustering methods to identify the features of the three multimode continuous processes previously presented in Chap. 2.

Chapter 4: **Monitoring of Multimode Continuous Processes** presents a methodology to design operating mode identification and fault detection methods for multimode processes. A fault can be considered as an unpermitted deviation from at least one characteristic property of the system which may interrupt the system's ability to perform a required function. An important challenge is to differentiate the intended operational change of mode from a fault. The complexity of this task is related to the types of modes of a process: steady, non-steady and mixed. Therefore, three different schemes for operating mode identification and fault detection of multimode processes are presented. Methods for fault detection during steady modes and transitions are also described. It is also illustrated how to apply these schemes for the monitoring of the cases of study presented in Chap. 2.

Chapter 5: **Fault Classification with Data-Driven Methods** presents a methodology to design fault cause identification methods based on data-driven classification tools. Once a fault has occurred in an industry, the measurements are usually archived. However, the data corresponding to the period where the fault was present can be used to determine in the future if the fault occurs again. This is achieved through the design of data-driven classification tools. The main difficulty of accomplishing such task in multimode processes is that the pattern of a fault might not be the same depending on the mode during which it occurs. Four classification tools are applied in this chapter for the fault cause identification of the three benchmarks previously studied. The challenges of the fault cause identification task are thus explored in this chapter.

Chapter 6: **Final remarks** highlights the main conclusions from the previous chapters as well as the future trends in monitoring multimode continuous processes with data driven methods.

Solving the Fault Diagnosis for Multimode Continuous Processes, through data-driven methods, brings to the readers' information from at least three different areas: Fault Diagnosis, Unsupervised and Supervised Learning, and Statistical Methods. Hence, the topic of the book covers a broad multi/interdisciplinary scientific community.

La Habana, Cuba   Marcos Quiñones-Grueiro
La Habana, Cuba   Orestes Llanes-Santiago
Nova Friburgo, Brazil   Antônio José Silva Neto
July 2020

Acknowledgements

The authors acknowledge the decisive support provided by CAPES- Foundation for the Coordination and Improvement of Higher Level Education Personnel, through the project "Computational Modelling for Applications in Engineering and Environment", Program for Institutional Internationalization CAPES PrInt 41/2017, Process No. 88887.311757/2018-00. Our gratitude is also due to the Springer Publisher, especially its representation in Germany, to CNPq- National Council for Scientific and Technological Development to FAPERJ - Foundation Carlos Chagas Filho for Research Support of the State of Rio de Janeiro, as well as to the Cuban Ministry of Higher Education MES (Ministerio de Educación Superior), Universidad Tecnológica de la Habana José Antonio Eceheverra, CUJAE and Universidade do Estado de Rio de Janeiro for the publication of this book. We would also like to acknowledge the important contribution of Professor Cristina Verde from Universidad Nacional Autónoma de México.

La Habana, Cuba   Marcos Quiñones-Grueiro
La Habana, Cuba   Orestes Llanes-Santiago
Nova Friburgo, Brazil   Antônio José Silva Neto
July, 2020

Contents

Acronyms and Nomenclature

Acronyms

CK	Calinski–Harabasz index
CRISP–DM	Cross-Industry Standard Process for Data Mining
CSTH	Continuous Stirred Tank Heater
CV	Cross-Validation
CVA	Canonical Variate Analysis
DB	Davies–Bouldin index
EM	Expectation-Maximization algorithm
FAR	False Alarm Rate
FCM	Fuzzy C-Means algorithm
FDR	Fault Detection Rate
FN	False Negative
FP	False Positive
GMM	Gaussian Mixture Models
HMM	Hidden Markov Models
HWDN	Hanoi Water Distribution Network
ICA	Independent Component Analysis
KDD	Knowledge Discovery from Data
KDE	Kernel Density Estimation
MDR	Missed Detection Rate
Min–Max	Scaling strategy for the variables
MM	Mixture Modeling clustering
N	Negative observations for binary classification problems
P	Positive observations for binary classification problems
PCA	Principal Component Analysis
PDF	Probability Density Function
PI	Plug-in
PLS	Partial Least Squares
ROC	Receiver Operating Characteristic

ROT	Rule of Thumb
Sil	Silhouette coefficient
SPE	Squared Prediction Error
SVDD	Support Vector Data Description
TEP	Tennessee Eastman Process
TN	True Negative observations for binary classification problems
TP	True Positive observations for binary classification problems
WB	Window-based clustering
WDN	Water Distribution Network

Nomenclature

b_i	Number of branches connected to node i of a water distribution network
$\mathbf{c}_i$	Centroid of cluster C_i
d_i	Demand of the consumers in node i of a water distribution network
$\widehat{d}_i$	Average Euclidean distance of all elements of cluster i to its centroid
$d(\mathbf{c}_i, \mathbf{c}_j)$	Euclidean distance between the centroids i and j
f_i	Leakage outflow in node i of a water distribution network
$f^D(x)$	Multivariate probability density function of the variable x given a data set D
$g(\mathbf{x}, \Omega)$	Discriminant function
k	Total number of clusters
h	Bandwidth/smoothness parameter for kernel density estimation
m	Number of variables
n	Number of observations
n_n	Number of observations of normal operation
n_f	Number of observations of operation with fault
n_h	Number of observations considered in one day
n_d	Number of days
n_o	Number of observations related to the operating modes
p_i	Pressure head in node i of a water distribution network
q_i	Flow in pipe i of a water distribution network
w_i	Prior probability of each component of a Gaussian mixture model
$\mathbf{w}_i$	Weights of a layer L_i in a neural network
x	Variable of a process

x^*	Scaled variable
$\mathbf{x}$	Vector with the set of variables of a process
A	Mixing matrix for independent component analysis
A_h	Probability transfer matrix of a Hidden Markov model
B_h	Observation probability matrix of a Hidden Markov model
C_i	Cluster i
$|C_i|$	number of observations in cluster C_i
$C = \{C_1, C_2, ..., C_k\}$	Set of clusters
D_n	Historical data set representing the nominal/normal/typical behavior of a process
D_f	Historical data set representing the faulty behavior of a process
D_i	Historical data set used for creating fault identification methods
D_o	Set of observations acquired online
$D = \{D_n, D_f\}$	Historical data set containing observations from both normal and faulty behavior of a process
$D_\Omega = \{D_{f_1}, D_{f_2}, ..., D_{f_z}\}$	Data set that represents the behavior of each fault
Dl	Statistical for fault detection during a transition
E	Residual subspace of input variables
F	Feature subspace of support vector data description
H_i	Subspace of vectors of weights in a neural network
I	Monitoring statistic for the independent component subspace
Ie	Monitoring statistic for the excluded independent component subspace
L_i	Layers of a neural network
O	Number of operating modes of a process
Q	Monitoring statistic for the residual subspace
S	Independent component subspace
$\mathbf{S}$	Set of states of a Hidden Markov model
$\mathbf{S_b}$	Between-classes scatter matrix
$\mathbf{S_w}$	Within-class scatter matrix
T^2	Monitoring statistic for the principal component subspace
V	Output function of a neural network
X	Data set of the measured variables of a process
$\tilde{X}$	Principal component subspace
X_{st}	Data segment of a transition
Y	Data set of the quality variables of a process
Z	Number of faults that affect a process
α	Threshold for fault detection during a transition

δ^{intra}	Average intra-cluster distance
δ^{inter}	Average inter-cluster distance
μ	Mean of a variable
ξ	Slack terms for support vector data description optimization
π	Initial state probability distribution of a Hidden Markov model
ρ_{ji}	Probability of observation $\mathbf{x}_j$ belong to cluster C_i
σ	Standard deviation of a variable
ϕ	Activation function in a neural network
Θ	Set of parameters of a Gaussian mixture model
Σ	Covariance matrix
Φ	Set of operating modes of a process
Φ_{svdd}	Non-linear transformation function of support vector data description
$\Omega = \{\omega_1, \omega_2, ..., \omega_Z\}$	Set of possible faults of a process
γ	Smoothing parameter in the kernel function

Chapter 1
Fault Diagnosis in Industrial Systems

Abstract This chapter presents the motivation for developing fault diagnosis application in industrial systems. Fault diagnosis methods can be broadly categorized into model-based and data-driven. Model-based strategies are briefly discussed while highlighting the increasing tendency to the use of data-driven methods given the increasing data available from process operations. The classic data-driven fault diagnosis loop is presented and each task is described in detail. A procedure is presented for the systematic design of data driven fault diagnosis methods. Finally, the fault diagnosis problem for multimode processes is briefly discussed.

1.1 Fault Diagnosis

There is a growing necessity to improve the operation performance of industrial processes. On one hand, greater efficiency in the usage of raw materials and higher level of quality in the final products allows increasing economic benefits. On the other hand, complying with environmental regulations and industrial security protocols is also of upmost importance [22, 44–46]. The occurrence of process failures has an adverse impact in the security of the operators, in the economic indicators, and possibly in the environment, translating into the non-compliance of operation specifications, the occurrence of unplanned operation stoppage and even industrial accidents with loss os lives. Because of these reasons, it is essential to incorporate in the supervision and control systems of industrial processes a fault diagnosis tool with the capability of detecting, identifying and isolating faults in the processes.

Fault diagnosis methods were presented at the beginning of the 1960–1970 decade. Those first approaches were based on monitoring some observable parameters of the processes, in order to verify if their values remained within the allowed operating ranges [23]. These methods were not useful for the early detection of faults which slowly affect the monitored parameters [23, 26].

Nowadays, strategies for fault diagnosis in industrial systems can be classified in two main groups. In the first group, the strategies to fault diagnosis are based on the use of a model of the process, obtained with the use of the first-principle laws or from a signal input-output relation. These strategies are known as model-based

© Springer Nature Switzerland AG 2021
M. Quiñones-Grueiro et al., *Monitoring Multimode Continuous Processes*,
Studies in Systems, Decision and Control 309,
https://doi.org/10.1007/978-3-030-54738-7_1

methods. In the second group, fault diagnosis methods are based on the use of a model obtained from historical data of the process, and they are, therefore, known as data-driven based methods [44–46].

This chapter presents in Sect. 1.2 a summary description of the main approaches to model-based fault diagnosis methods. In Sect. 1.3, data-driven fault diagnosis methods are presented in more detail.

1.2 Model-Based Fault Diagnosis Methods

There are three main approaches to model-based fault diagnosis: methods based on state observers, parity spaces, and parameter estimation. The fundamental principle of these methods is the generation of a vector of residuals resulting from the difference between the measured variables of the process and the variables estimated from the model [21, 23, 25, 31, 33, 36].

1.2.1 State Observers

The methods based in state observers were introduced for linear systems in the 1970–1980 decade [3, 12, 32, 47].

These methods reconstruct the states of the system by using the measurements made in the system inputs and outputs. The difference between the measured outputs and the estimated outputs is used as the vector of residuals. They require a detailed mathematical model of the plant, preferably derived from first principles such that the states in the state-space equations have a physical interpretation.

Some examples of the use of this technique can be found in [18, 29, 34, 35, 53].

1.2.2 Parity Space

Different approaches of the Parity Space methods were presented in the 1980–1990 decade [11, 15, 51]. They were formally established in [38].

These methods perform a consistency check of the mathematical equations of the system with the measurements [16]. The advantage of this approach is that, given a transfer function, it is possible to design the residual generator based on the parity space, without the computational load and the knowledge involved in the realization of the state space.

There is a close relationship among the approaches of state observers and parity spaces. Some papers which show such relationships are [9, 13, 14, 28, 42, 44, 48]. Strategies have been developed with the joint use of both approaches where the system design is carried out leveraging from the computational advantages of the parity space approach while the solution is carried out in the form of observers to ensure numerical stability and a lower online computational load [40].

1.2.3 Parameter Estimation

Parameter estimation methods have been widely used for monitoring processes where non directly measurable quantities must be estimated. Residuals are obtained from the difference between the parameters of the nominal model and parameters estimated based on the process measurements. Deviations in model parameters serve as a basis for detecting and isolating faults. In [23, 24], the details of several methods are presented.

In recent years, a new approach has been presented within parameter estimation methods based on the solution of an inverse problem. This approach has shown excellent robustness characteristics in the presence of noise and external disturbances, as well as high sensitivity to faults of small magnitude [5–8].

Modern industries have a great complexity due to the increasing development of electronics, industrial instrumentation, and computation. Then, obtaining models that allow the use of the techniques described above is a difficult task. That is the reason why in recent years fault diagnosis techniques based on historical data are being adopted in most industries.

1.3 Data-Driven Based Fault Diagnosis Methods

In contrast to the model-based approach where prior knowledge of the process is needed, data-driven methods only require the availability of a large amount of data to create fault diagnosis models. Data-driven analysis is a discipline which has gained popularity in recent years thanks to the rapid development of sensor and communication technologies. Being possible to collect in real time high volumes of data from industrial processes, and to storage large data sets in remote servers, the challenge nowadays is to extract useful information to build better fault diagnosis methods. The process of automating the extraction of information from data is known as Data Mining, also referred to as knowledge discovery from data (KDD) [19, 39]. In this section, the general approach to data-driven fault diagnosis is first explained. Then, a methodology for the design of data-driven fault diagnosis methods is presented, based on a well known data mining methodology.

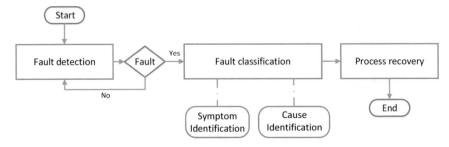

Fig. 1.1 Classic fault diagnosis/monitoring loop

1.3.1 Data-Driven Fault Diagnosis/monitoring Loop

The data-driven fault diagnosis/monitoring loop traditionally considers three main tasks depicted in Fig. 1.1: fault detection, fault classification, and process recovery. Each task is presented next.

1.3.1.1 Fault Detection

The primary fault diagnosis task is the fault detection, which can be described as the evaluation of the process current behavior in order to determine if it is normal/typical or not. The effectiveness of the data-driven fault diagnosis approach will ultimately depend on the adequate characterization of the data in order to find behavior patterns. There are two causes for the variations in the variables of a process: common causes and special causes [37]. Common causes of variation occur only due to random noise, while special causes are all those variations which may not be attributable to random noise.

Standard control strategies of industrial processes may be able to compensate for many of the special causes of variation, but not for common causes, which are inherent to the process data. Because variation in the data is inevitable, statistical theory plays a fundamental role in many process diagnosis schemes. The application of statistical theory for fault diagnosis of processes is based on the assumption that the characteristics of the data variations remain relatively unchanged until a system fault occurs. This implies that properties of the data such as the mean and variance of each variable are repeatable given the same operating condition of the process. This repeatability of the statistical properties allows to establish the characteristics that define the normal operating behavior or state. Consequently, fault detection can be stated as the problem of detecting changes in the statistical properties of the data representing the normal operating mode of a process. Therefore, the fault detection task can be defined as follows

Definition 1.1 (*Data-driven fault detection task*) Given a data set formed by $n = n_n + n_f$ observations in a m dimensional space ($D = \{\mathbf{x}\}_{j=1}^{n}, \mathbf{x}_j \in \Re^m$, $D = D_n +$

D_f) where n_n are a set of observations representative of the normal operation of the process and n_f are a set of observations representative of the faulty operation of the process, find a function $f_d(\mathbf{x})\colon \Re^m \rightarrow [0, 1]$ that maps any observation of the feature space to a set of two possible scenarios 0: normal and 1: fault.

The most important requirement to design a fault detection function is to have a data set representative of the process normal operation. Otherwise, typical patterns of the data may be wrongly identified as faults when the detection function $f_d(\mathbf{x})$ is used. In Ref. [2] the design of online change detection algorithms and an analysis of their performances is presented. Both Bayesian and non-Bayesian approaches are discussed.

1.3.1.2   Fault Classification

Once detected, there is an interest on identifying the cause of the fault, as well as the component/part of the process which is failing. Therefore, the classification task can be divided into two sub-tasks: the variable/symptom identification of the fault, and the recognition of the cause of the fault. The first sub-task is defined as follows

Definition 1.2 (*Data-driven fault symptom identification*) Given a data set formed by n_f observations in a m dimensional space ($D_f = \{\mathbf{x}\}_{j=1}^{n_f}, \mathbf{x}_j \in \Re^m$, $D_f \subset D$) that is representative of a fault that has been detected, and the set of m variables $X = \{x_1, x_2, ..., x_m\}$ of the process, find a function $f_i(D_f, X)\colon D_f \in \Re^{n_f \times m} \rightarrow \hat{X} = \{x_1, ..., x_p\}$ that maps the fault data set to a subset of $p < m$ variables $\hat{X} \subset X$ associated with the cause of the fault.

The data-driven fault symptom identification subtask is also known as fault identification [10]. The information provided by a fault symptom identification system is helpful for the operator to focus his attention on the most likely physical component(s)/part(s) of the process which are failing [27]. Many statistical and artificial intelligence techniques have been developed to solve this task [1, 41].

The second sub-task is defined as follows

Definition 1.3 (*Data-driven fault cause recognition*) Given a data set formed by n_f observations in a m dimensional space ($D_f = \{\mathbf{x}\}_{j=1}^{n_f}, \mathbf{x}_j \in \Re^m$, $D_f \subset D$) that is representative of a fault that has been detected, and a set of possible causes $\Omega = \{\omega_1, \omega_2, ..., \omega_Z\}$ find a function $f_c(D_f, \Omega)\colon D_f \in \Re^{n \times m} \rightarrow \omega_z \in \Omega$ that associates the data set D_f with one of the Z possible causes.

The fault cause recognition sub-task is usually formulated as a pattern classification problem when data from faults that have occurred previously are available to create the function $f_c(D_f, \Omega)$. In this case, the pattern associated with each fault is known as a class, and the decision function should try to guarantee the best possible isolability among the classes by following the *learning from examples* paradigm. The *learning from examples* paradigm establishes that the parameters of the function

$f_c(D_f, \Omega)$ can be estimated as long as some objects (observations) from the classes population are available [20]. In the context of machine learning methods, the fault classification based on fault data is framed as a supervised learning problem because the examples of each class are known (the classes are labeled).

If these two sub-tasks are solved, the fault's severity can be identified, which is extremely important in many industries to make decisions about the type of process recovery to be adopted.

1.3.1.3 Process Recovery

The final major task of the fault diagnosis/monitoring loop is the process recovery. This task consists of selecting the correcting action(s) or routine to be followed by the operator to remove/compensate the effect of the fault on the operation of the system. The automation of this task is not always feasible depending on the complexity of the fault, the cost of the correction action, and the type of process [30]. Therefore, once a dangerous fault is diagnosed, the most common solution to avoid accidents is to stop the process operation.

1.3.2 Design of the Data-Driven Fault Diagnosis/monitoring Loop

Fault diagnosis based on data can be framed as a data mining problem. The most popular methodology to develop data mining applications is the Cross Industry Standard Process for Data Mining (CRISP-DM) because of its stepwise procedure and general applicability [4]. Such methodology can be adapted to the design of data driven fault diagnosis tasks. The lifecycle of a data mining project can be described by the CRISP-DM methodology model shown in Fig. 1.2 [49]. Following, the CRISP-DM methodology applied to the design of each stage of the fault monitoring loop is presented.

1.3.2.1 Data-Driven Fault Detection Task

- *Step 1*
  The first step (business understanding) is related to understanding the objectives and requirements of the business as well as whether data mining can be applied to meet them. In this step, the project manager must also decide what type of data should be collected to build a useful model. For fault detection, this step is related to:

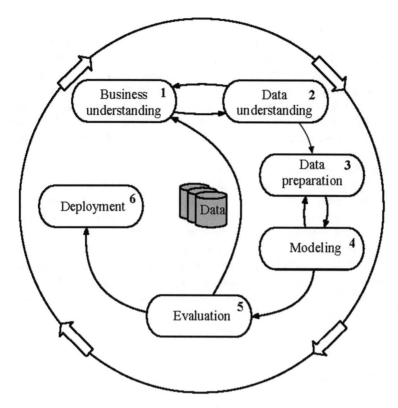

Fig. 1.2 Lifecycle of the Cross Industry Standard Process for Data Mining (CRISP-DM) [49]

1. Selecting the process variables that will be used to design the fault detection function $f_d(\mathbf{x})$.
2. Selecting a data set D representative of the normal/typical operation of a process.
3. Characterizing the features of the process which will be considered to design the fault detection function to be used.

The features of the process to be analyzed are

1. Linear or non-linear process. Are the relationships among the variables of the process linear or non-linear?
2. Dynamic or not. Are the variables of the process sampled at a rate that allows to characterize the dynamic behavior of the process with sufficient accuracy?
3. Continuous or batch. Is the process operation continuous or in batch?
4. Multimode or not. The process operates under multiple operating conditions or not?

- *Step 2*

  In step two (data understanding), the quality of the initial data set (D) is studied to verify whether it is suitable for further processing. Analysis of the data at this step may help to re-evaluate the feasibility of using data mining to achieve the business goals. For fault detection this step involves the verification of the presumed features of the process based on the data driven properties of the data set. Since this book focuses on multimode continuous processes, a method to verify the multimode behavior of the process from a data set will be introduced in the next chapter.

- *Steps 3 and 4*

  The next two steps, data preparation and modeling, are the main focus of this book. Preparation involves pre-processing the raw data to be used for building the fault detection function. Modeling involves using a method to design the fault detection function based on both the properties of the process and the data set. There are different ways in which data can be transformed, and presented as prior knowledge to a diagnostic system, which is known as knowledge extraction [17]. Data preparation and modeling usually are co-dependent. Most of the times, it is necessary to iterate: results obtained during modeling provide new insights that affect the choice of pre-processing techniques. Different data preparation techniques for fault detection will be covered in subsequent chapters of this book. The appropriate fault detection model depends to a great extent on the proper pre-processing of the data.

- *Step 5*

  This step is fundamental for the success of the data mining application: evaluation. The use of appropriate evaluation measures allows to judge the effectiveness of the proposed approach. The evaluation measures to consider for the fault detection task will be presented in the Sect. 1.3.3.

- *Step 6*

  The final step (deployment) usually involves the integration of the fault diagnosis application into a larger software system. This is the stage where the implementation details of the data mining techniques matter. It may be required, for instance, to re-implement the application in a different programming language.

1.3.2.2 Data-Driven Fault Symptom Identification Sub-task

- *Step 1*

  The first step is related to:

1. Selecting the process variables that will be used to design the function $f_i(D_f, X)$.
2. Selecting a data set $D_i \subset D$ that will be used to build the function.
3. Characterizing the features of the process. The features to be analyzed are the ones previously mentioned.

Note in this case that the data set D_i which will be used to create f_i may not necessarily be a data set containing faults. Overall, such data set must allow the characterization of the process behavior, such that when a fault occurs the variables causing the fault can be identified. Therefore, D_i may contain both data of the normal and faulty behaviors, which will be used to calibrate the parameters of the symptom identification function.

- *Step 2*
  Similar to the fault detection task, the step two (data understanding) for fault symptom identification involves the verification of the features of the process based on the data driven properties of the data set, i.e. the study of the data quality.

- *Steps 3 and 4*
  The next two steps, data preparation and modeling, are also similar to the ones in the fault detection task. These are related to pre-processing the raw data, and using a method to design the symptom identification function.

- *Steps 5 and 6*
  These steps are similar to the ones in the fault detection task but the evaluation measures to be considered are different. This difference is given by the fact that the data driven problem to be solved for the fault symptom identification task is different from the one in the fault detection task. Specifically, the fault detection task is framed as a binary classification problem, while the fault symptom identification task is framed as a multi-class classification problem. The evaluation measures will be presented in the Sect. 1.3.3.

1.3.2.3   Fault Cause Identification Sub-task

- *Step 1*
  The first step is related to:

1. Selecting the process variables that will be used to design the function $f_c(D_f, \Omega)$.
2. Selecting a data set that represents the behavior of each fault in $D_\Omega = \{D_{f_1}, D_{f_2}, ..., D_{f_z}\}$.
3. Characterizing the features of the process. The features to be analyzed are the same previously mentioned for the fault detection task.

Note that in this case, a data set representative of each fault should be available.

- *Steps 2–6*
  The remaining steps are similar to those associated with the fault symptom identification sub-task. The evaluation measures used are also the same because the fault cause identification sub-task is also framed as a multi-class classification problem.

1.3.3 Evaluation Measures Adopted to Design Fault Diagnosis Methods

Evaluation measures are employed to test the quality of a data-driven method for a particular process and fault diagnosis. A data-driven method selected to achieve any of the three fault diagnosis tasks, i.e. fault detection, fault classification, and process recovery, might present a good performance for some process but bad performance for others. This may be considered as the No Free Lunch theorem [50] for fault diagnosis applications: there is no universal data-driven method that allows successful fault diagnosis for any process. This implies that the use of evaluation measures to calibrate and select the appropriate method for each fault diagnosis task is fundamental.

- **Evaluation measures for fault detection**
  The fault detection task can be framed as a binary classification problem. Therefore, all evaluation indices for fault detection are obtained from a confusion matrix, which is represented in Table 1.1 [19].
  A positive observation is the one obtained when a fault occurs in the process. Conversely, a negative observation is obtained when the behavior of the process is normal. P and N refer here to the true number of positive and negative observations respectively in a data set $D = \{D_n, D_f\}$ that contains observations representative of the normal and faulty behavior. P' and N' are the observations classified as positive and negative respectively, TP (True positives) refers to the positive observations which are correctly labeled, TN (True negatives) are the negative observations which are correctly labeled, FP (False positives or Error type 1) are the negative observations which are incorrectly labeled as positive, and FN (False negatives or Error type 2) are the positive observations which are incorrectly labeled as negative.
  Given a data set that represents the normal behavior of the process, D_n, and another one obtained when faults occur, D_f, the fault detection function should be calibrated to guarantee:

1. A low false alarm rate: $FAR = FP/N$. The reliability of the fault diagnosis application mainly depends on the false alarm rate. A high FAR would harm the confidence of the process operators in the fault diagnosis application. Furthermore, in some industries a false alarm may cause economic losses because of unnecessary process stoppages. Therefore, a rule of thumb for the calibration of fault detection methods is to keep a FAR lower than 5% at the design stage.
2. A small fault latency or fault detection delay. Once a fault occurs, the time frame during which observations representing a fault are wrongly classified as normal defines the fault latency. The detection delay should be taken into account especially for critical systems. If there is a set of faults affecting the process such that they can cause accidents or significant economic losses, then their early detection should be considered of primary importance.
3. A low missed detection rate: $MDR = FN/P$. The existence of a fault should be verified once it has been detected by a persistence indicator: the MDR. Note

Table 1.1 Confusion matrix

		Predicted class		Total
		Yes	No	
Actual class	Yes	TP	FN	P
	No	FP	TN	N
	Total	P′	N′	P+N

that if the FAR is guaranteed to be very low, then the process operator can assure the existence of a fault immediately after it has been detected.

Other performance indicator for binary classification problems is the Receiver Operating Characteristic plot or ROC curve [52]. This graph shows the trade-off between true positives and true negatives. The larger the area under the curve, the better is the monitoring model performance.

- **Evaluation measures for fault symptom identification and fault cause identification**

  These two fault diagnosis sub-task can be analyzed as a multi-class pattern recognition problem. In such case, a confusion matrix is defined as $A = [A(i, j)] \in \mathfrak{R}^{Z \times Z}$ where Z is the number of faults (classes) whose symptoms or causes should be identified. For the symptom identification sub-task each subset of variables is associated with a class. For the cause identification sub-task, the data set acquired during the occurrence of a specific fault is associated with a class. The confusion matrix is formed such that each element $A(i, j)$ contains the number of observations with a fault label i that are classified as fault label j [43]. A fault symptom or cause identification function should be calibrated to guarantee:

1. A high overall accuracy: $Ac = \frac{1}{n} \sum_{i=1}^{z} A(i, i)$ where n_f is the total number of observations from all faults. The overall accuracy represents the fraction of observations which have been correctly diagnosed. The overall percentage error is $Err = 100(1 - Ac)\%$.

1.3.4 Data-Driven Fault Diagnosis of Multimode Continuous Processes

The focus of this book is the design of data driven fault detection strategies for multimode continuous processes. The features of this type of processes requires the combination of different tools to achieve a satisfactory performance. The fault classification and process recovery tasks are not presented in this book because these tasks are solved once a fault and an operating mode have been identified. Therefore, the fault detection task plays a more fundamental role because a fault must be distinguished from a intentional programmed change of operating mode.

In step one of the CRISP-DM methodology for fault detection one decides if the process is multimode or not from a practical point of view. Since the expert knowledge is not always available to decide in advance if a process is multimode or not, a data-driven method can be used to verify the multimode behavior in step two of the CRISP-DM methodology. Therefore, in Chap. 2 a multimode process from a data-driven perspective is mathematically defined. Moreover, a method is presented to determine from a data set if a process is considered as multimode or not. Three benchmark cases of study which will be used throughout the book are also presented.

Remarks

In this chapter, the two main fault diagnosis approaches for industrial systems are presented: model-based and data-driven methods. The different tasks of the data driven fault diagnosis loop are defined and explained. A methodology to design the different fault diagnosis tasks based on the CRISP-DM methodology for data mining applications was presented step by step. This methodology allows to systematically design fault diagnosis tasks.

References

1. Alcala, C.F., Qin, S.J.: Reconstruction-based contribution for process monitoring. Automatica **45**, 1593–1600 (2009)
2. Basseville, M., Nikiforov, I.V.: Detection of Abrupt Changes: Theory and Application. Prentice Prentice-Hall, Inc. (1993)
3. Beard, R.V.: Failure accomodation in linear systems through self-reorganization. Ph.D. thesis, MIT (1971)
4. Chapman, P., Clinton, J., Kerber, R., Khabaza, T., Reinartz, T., Shearer, C., Wirth, R.: CRISP-DM 1.0 - Step-by-step data mining guide. CRISP-DM Consortium (2000). https://the-modeling-agency.com/crisp-dm.pdf. Accessed on May 2020
5. Camps Echevarría, L., Silva Neto, A.J., Llanes-Santiago, O., Hernández Fajardo, J.A., Jiménez Sánchez, D.: A variant of the particle swarm optimization for the improvement of fault diagnosis in industrial systems via faults estimation. Eng. Appl. Artif. Intell. **28**, 36–51 (2014)
6. Camps Echevarría, L., Campos Velho, H.F., Becceneri, J.C., Silva Neto, A.J., Llanes-Santiago, O.: The fault diagnosis inverse problem with ant colony optimization and ant colony optimization with dispersion. Appl. Math. Comput. **227**(15), 687–700 (2014)
7. Camps Echevarría, L., Llanes-Santiago, O., Fraga de Campos Velho, H., Silva Neto, A.J.: Fault Diagnosis Inverse Problems: Solution with Metaheuristics. Springer (2019). https://doi.org/10.1007/978-3-319-89978-7
8. Camps Echevarría, L., Llanes-Santiago, O., Silva Neto, A.: An approach for fault diagnosis based on bio-inspired strategies. In: IEEE Congress on Evolutionary Computation. IEEE, Barcelona, Spain (2010)
9. Chen, J., Patton, R.J.: Robust Model-based Fault Diagnosis for Dynamic Systems. Kluwer Academic Publishers, Dordrecht (1999)
10. Chiang, L.H., Russell, E.L., Braatz, R.D.: Fault Detection and Diagnosis in Industrial Systems. Springer (2001)
11. Chow, E.Y., Willsky, A.: Analytical redundancy and the design of robust failure detection systems. IEEE Trans. Autom. Control **29**, 603–614 (1984)
12. Clark, R.: Instrument fault detection. IEEE Trans. Aero. Electron. Syst. **AES14**, 456–465 (1978)

13. Ding, S.X.: Model-based Fault Diagnosis Techniques: Design Schemes, Algorithms, and Tools. Springer (2008)
14. Florin Metenidin, M., Witczak, M., Korbicz, J.: A novel genetic programming approach to nonlinear system modelling: application to the DAMADICS benchmark problem. Eng. Appl. Artif. Intell. **17**(4), 363–370 (2004)
15. Gertler, J.: A numerical-structure approach to failure detection and isolation in complex plants. In: Proceedings of the 25th CDC, pp. 1576–1580 (1986)
16. Gertler, J.: Fault detection and isolation using parity relations. Control Eng. Pract. **5**(5), 653–661 (1997)
17. Gertocio, C., Dussauchoy, A.: Knowledge discovery from industrial databases. J. Intell. Manuf. **15**(1), 29–37 (2004)
18. Guo, J., Yuan, T., Li, Y.: Fault detection of multimode process based in local neighbor normalized matrix. Chemom. Intell. Lab. Syst. **154**, 162–175 (2016)
19. Han, J., Kamber, M., Pei, J.: Data Mining Concepts and Techniques. Elsevier (2012)
20. Heijden, F.V.D., Duin, R., Ridder, D.D., Tax, D.: Classification, Parameter Estimation and State Estimation. Wiley (2004)
21. Hoefling, T., Isermann, R.: Fault detection based on adaptive parity equations and single-parameter tracking. Control Eng. Pract. **4**(10), 1361–1369 (1996)
22. Hwang, I., Kim, S., Kim, Y., Eng, C.: A survey of fault detection, isolation, and reconfiguration methods. IEEE Trans. Control Syst. Technol. **18**(3), 636–653 (2010)
23. Isermann, R.: Process fault detection based on modelling and estimation methods-a survey. Automatica **20**(4), 387–404 (1984). https://doi.org/10.1016/0005-1098(84)90098-0
24. Isermann, R.: System Fault Diagnostics, Reliability and Related Knowledge-Based Approaches, Chapter Experiences with Process Fault Detection Methods via Parameter Estimation, pp. 3–33. Springer, Dordrecht (1987)
25. Isermann, R.: Fault diagnosis of machines via parameter estimation and knowledge processing. Automatica **29**(4), 815–835 (1993)
26. Isermann, R.: Supervision, fault-detection and fault-diagnosis methods-an introduction. Control Eng. Pract. **5**(5) (1997)
27. Isermann, R.: Model-based fault-detection and diagnosis-status and applications. Annu. Rev. Control **29**, 71–85 (2005)
28. Isermann, R.: Model based fault detection and diagnosis. Status and applications. Annu. Rev. Control **29**(1), 71–85 (2005)
29. Isermann, R.: Fault Detection Systems, Chapter Fault Detection with State Observers and State Estimation, pp. 231–252. Springer, Berlin (2006)
30. Isermann, R.: Fault-Diagnosis Applications. Springer (2011)
31. Isermann, R., Ballé, P.: Trends in the application of model-based fault detection and diagnosis of technical processes. Control Eng. Pract. **5**(5), 709–719 (1997)
32. Jones, H.: Failure detection in linear systems. Ph.D. thesis, MIT (1973)
33. Krishnaswami, V., Luh, G.C., Rizzoni, G.: Nonlinear parity equation based residual generation for diagnosis of automotive engine faults. Control Eng. Pract. **3**(10), 1385–1392 (1995)
34. Methnani, S., Gauthier, J., Lafont, F.: Sensor fault reconstruction and observability for unknown inputs, with an application to wastewater treatment plants. Int. J. Control **84**(4), 822–833 (2011)
35. Mohamadi, L., Dai, X., Busawon, K., Djemai, M.: Output observer for fault detection in linear systems. In: 2016 IEEE 14th International Conference on Industrial Informatics (INDIN), Poitiers, France (2016)
36. Odgaard, P.F., Matajib, B.: Observer-based fault detection and moisture estimating in coal mills. Control Eng. Pract. **16**, 909–921 (2008)
37. Ogunnaike, B.: Process Dynamics, Modeling, and Control. Oxfor University Press, New York (1994)
38. Patton, R.J., Chen, J.: A review of parity space approaches to fault diagnosis. In: IFAC SAFE-PROCESS Symposium (1991)
39. Prieto-Moreno, A., Llanes-Santiago, O., García Moreno, E.: Principal components selection for dimensionality reduction using discriminant information applied to fault diagnosis. J. Process Control **33**, 14–24 (2015)

40. Schneider, S., Weinhold, N., Ding, S., Rehm, A.: Parity space based FDI-scheme for vehicle lateral dynamics. In: Proceedings of 2005 IEEE Conference on Control Applications, CCA 2005. Toronto, Ont., Canada (2005)

41. Shang, L., Liu, J., Zhang, Y.: Recursive fault detection and identification for time-varying processes. Ind. Eng. Chem. Res. **55**(46), 12149–12160 (2016)

42. Simani, S., Fantuzzi, C., Patton, R.J.: Model-Based Fault Diagnosis in Dynamic Systems Using Identification Techniques. Springer (2002)

43. Theodoridis, S., Koutroumbas, K.: Pattern Recognition. Elsevier (2009)

44. Venkatasubramanian, V., Rengaswamy, R., Yin, K., Kavuri, S.N.: A review of process fault detection and diagnosis-Part I: quantitative model-based methods. Comput. Chem. Eng. **27**(3), 293–311 (2003). https://doi.org/10.1016/s0098-1354(02)00161-8

45. Venkatasubramanian, V., Rengaswamy, R., Yin, K., Kavuri, S.N.: A review of process fault detection and diagnosis-Part II: qualitative model-based methods and search strategies. Comput. Chem. Eng. **27**(3), 313–326 (2003). https://doi.org/10.1016/s0098-1354(02)00161-8

46. Venkatasubramanian, V., Rengaswamy, R., Yin, K., Kavuri, S.N.: A review of process fault detection and diagnosis-Part III: process history based methods. Comput. Chem. Eng. **27**(3), 327–346 (2003). https://doi.org/10.1016/s0098-1354(02)00161-8

47. Willsky, A.S.: A survey of design methods for failure detection systems. Automatica **12**(6), 601–611 (1976)

48. Witczak, M.: Modelling and Estimation Strategies for Fault Diagnosis of Non-Linear Systems From Analytical to Soft Computing Approaches, vol. 354. Springer (2007). https://doi.org/10.1007/978-3-540-71116-2

49. Witten, I.H., Frank, E., Hall, M.A., Pal, C.J.: Data Mining: Practical Machine Learning Tools and Techniques. Morgan Kaufmann (2016)

50. Wolpert, D.H., Macready, W.G.: No free lunch theorems for optimization. IEEE Trans. Evol. Comput. **1**(1), 67–82 (1997)

51. Wünnenberg, J., Frank, P.: Model-based residual generation for dynamics systems with unknown inputs. In: Proc. 12th IMACS World Congress on Scientific Computation, Paris, vol. 2, pp. 435–437 (1988)

52. Zaki, M.J., Meira, W.: Data Mining and Analysis. Cambridge University Press (2014)

53. Zhu, F., Cen, C.: Full-order observer-based actuator fault detection and reduced-order observer-based fault reconstruction for a class of uncertain nonlinear systems. J. Process Control **20**(10), 1141–1149 (2010)

Chapter 2
Multimode Continuous Processes

Abstract This chapter presents the characteristics and definitions for multimode continuous processes. It is discussed why they require specific monitoring approaches for achieving good fault sensitivity and low false alarm rate. The features of a multimode continuous process are formalized mathematically from a data-driven point of view. Moreover, the different types of multimode processes of practical interest are presented and illustrated.

2.1 Features of Multimode Continuous Processes

An industrial process is generally presented as multimode if the operating conditions are not constant [11, 15, 17, 31, 33–35]. The existence of multiple modes is caused by several factors. For instance, in the manufacturing industry, variations in the desired production volume and feedstock material alterations will generally occur throughout the year [10]. In the water industry, water consumption habits vary according to the season of the year, e.g. summer-winter, weekdays-weekend, holidays, etc. [19]. In thermal power plants, the consumers' demand for energy changes daily [29]. Even normal maintenance tasks may affect the operating conditions of refining processes [1].

The changes of operating conditions are reflected in the features of the variables that characterize the process [36, 37]. If there exists a narrow range of variation that remains constant for each variable for a given time period, then it is usually considered that the process operates in a steady mode. The steady mode condition is formulated mathematically as follows [3, 26]

$$\left| \frac{x(t) - x(t_0)}{t - t_0} \right| < T_x \ \forall t \in [t_0, t_0 + \Delta t] \tag{2.1}$$

where $x(t)$ is the value of any variable at time t, $x(t_0)$ represents the value of the same variable at initial time t_0, and T_x is a threshold which is adjusted depending on both measurement units and the noise variance for the specific variable. There are numerous ways to tune T_x for each variable [14]. When a mode change occurs,

© Springer Nature Switzerland AG 2021
M. Quiñones-Grueiro et al., *Monitoring Multimode Continuous Processes*,
Studies in Systems, Decision and Control 309,
https://doi.org/10.1007/978-3-030-54738-7_2

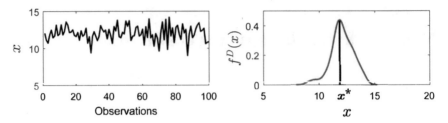

Fig. 2.1 Example of variable showing a steady mode process

the process usually undergoes a transition mode such that at least one variable will violate the narrow range of variation given by T_x.

The main limitation in the use of the above method to define whether a process is multimode or not cosist in the need of a priori knowledge, i.e.

1. Selecting the parameter T_x of each variable.
2. Selecting the time interval to analyze Δt. The behavior of the processes depends on physical and chemical phenomena. This implies that the duration and nature of the transitional modes may vary significantly, depending on the type of process.

The duration of a transition mode depends on the dynamic behavior of the process. Chemical processes within the chemical and pharmaceutical industry generally present slow dynamics such that the transition mode can last for long periods of time compared to the sampling time of the process (i.e. hours). On the other hand, transitions in the energy industry usually last very short periods of time (i.e. minutes). Transitions can also occur off-line. For instance, a manufacturing plant can be turned off before modifying the features of the product to be fabricated.

Following, it is presented a formal definition of the steady and transition modes based on the statistical properties of the sampled observations, by considering an outlier-free data set, which does not require a prior knowledge of the process. These two definitions also allow to establish a formal definition of a multimode continuous process.

Definition 2.1 (*Steady mode for continuous processes*) A data set $D=\{\mathbf{x}_1, \mathbf{x}_2, ..., \mathbf{x}_n\}$ of n continuously sampled observations, where $\mathbf{x}_j \in \mathfrak{R}^m$ and a multivariate probability density function of a variable denoted by $f^D(x)$, represent a steady mode if the following condition is satisfied:

- *For all variables of $\mathbf{x}_j$ there exists only one local maximum of $f^D(x)$:* $\arg\max\limits_{x} f^D(x) = S$, *where the set of solutions has a cardinality equal to 1* $(|S| = 1)$ *for all variables in D.*

From the definition given above, an operating mode is considered steady if the multivariate probability density function (PDF) of any variable in the data set has only one local maximum. An academic univariate example of a variable with a single maximum at $S = \{x^*\} = 11.93$ is shown in Fig. 2.1.

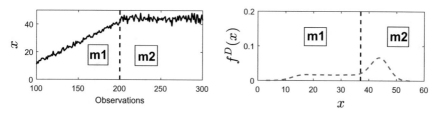

Fig. 2.2 Example of a variable showing a transitional (1) and a steady mode (2)

Definition 2.2 (*Transitional mode for continuous processes*) A data set $D = \{x_1, x_2, ..., x_n\}$ of n continuously sampled observations, where $x_j \in \Re^m$, and a multivariate probability density function of a variable denoted by $f^D(x)$, represent a transitional mode if the following condition is satisfied:

- *There exists more than one local maximum of* $f^D(x)$ *for any of the variables in* x_j: $\arg\max\limits_x f^D(x) = S$ *and* $|S| > 1 \; \forall x \in x_j$.

An academic univariate example of a variable showing a transitional (m1) and a steady mode (m2) is shown in Fig. 2.2. The vertical dashed line illustrates when the transition ends.

Once a steady and a transitional mode have been defined, a process can be characterized as multimode if it presents at least two operating modes as described next.

Definition 2.3 (*Multimode continuous process*) A data set $D = \{x_1, x_2, ..., x_n\}$ of n continuously sampled observations, where $x \in \Re^m$, and a multivariate probability density function of a variable denoted by $f^D(x)$, represent a multimode process if there exists at least two nonempty and disjoint subsets of D formed by continuously sampled observations ($C_1 \neq \emptyset$, $C_2 \neq \emptyset$, $C_1 \cap C_2 = \emptyset$, $C_1 \subset D$, and $C_2 \subset D$) representing two different types of operating modes, or two different steady operating modes.

It is necessary to comment that although the mode types of a multimode process could be defined from a statistical perspective as stationary or non-stationary, the definitions of a strict or wide stationary process [20] are hard to meet within the industrial context. From the perspective of such formal definitions, industrial processes are usually operating in non-stationary conditions.

2.2 Multimode Process Assessment Algorithm

The definitions of steady and transition modes do not allow to determine directly from a data set, the number of operating modes of a process. However, they allow to determine if a data set represents a multimode process. Therefore, this section presents an algorithm that allows to automatically estimate the number of maximums of the

Fig. 2.3 Probability density
function of a continuous
variable

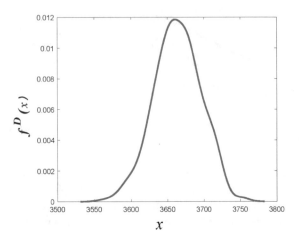

probability density function of a variable. The use of the algorithm demonstrates the applicability of the presented definitions from a practical standpoint.

2.2.1   Preliminary Theory

The number of maxima of the probability density function of a variable in a data set free of outliers determine if the process is multimode or not. Therefore, the probability density function (PDF) of a continuous random variable is described first.

The PDF of a variable provides the likelihood that the variable takes any given value. See for example Fig. 2.3 where it is more likely that the variable takes the value of 3650 than the value of 3550, given the PDF. Moreover, the probability that the variable takes the value of 3500 is almost zero.

To calculate the maximum of the PDF, it is required to estimate the shape of the PDF. Kernel Density Estimation (KDE) methods are mathematical tools that can be used for such purpose [24]. The idea behind KDE is that the influence of each observation in the PDF of a variable can be modeled formally by using a mathematical function called Kernel. In other words, the Kernel function describes the influence of a data point within its neighborhood. The formal definitions of Kernel function, continuous PDF, and PDF discrete estimate are presented next.

Definition 2.4 (*Kernel function*) A kernel function is a function $K: \mathfrak{R}^m \to \mathfrak{R}$, $K(\mathbf{x}) \geq 0$ which has the following property: $\int_{\mathfrak{R}^m} K(\mathbf{x})d\mathbf{x} = 1$.

Definition 2.5 (*Continuous probability density function*) $f(x)$ is a probability density function of a continuous variable x if the following conditions are satisfied:

- $f(x)$ is non-negative: $f(x) \geq 0 \ \forall x$.
- $f(x)$ must integrate to one: $\int_{-\infty}^{+\infty} f(x) = 1$.

Definition 2.6 (*Probability density function discrete estimate*) Given a set of n observations of a variable x_l ($L = \{x_l\}_{l=1}^n, x_l \in \Re$) the estimate of the discrete probability density function is defined as

$$\hat{f}^L(x) = \frac{1}{nh} \sum_{l=1}^n K(\frac{1}{h}(x - x_l)).$$

where K is a Kernel function and h is a smoothness parameter usually known as bandwidth.

The last definition implies that the evaluation of the discrete probability density function at one observation equals approximately the sum of the kernels of all observations. It is considered that the differences in the estimation error between various kernel types are not significant. Therefore, the choice of bandwidth h has a bigger impact in the PDF estimation than the choice of the kernel. Because the Gaussian or normal kernel generates the smoothest density curves, it is the most often used in practice [8]. Therefore, the Gaussian kernel will be adopted in this book.

2.2.2 Algorithm

The algorithm used to estimate the maxima of the discrete PDF is known as the mean-shift algorithm [8]. It consists on a gradient-based optimization approach with respect to the PDF. The definition of the PDF discrete estimate based on a Gaussian kernel $\hat{f}_G^D(x)$ and its gradient $\nabla \hat{f}_G^D(x)$ are presented below.

Definition 2.7 (*Gaussian Kernel function*) For a variable $x \in \Re$ the Gaussian kernel function is defined as $K_G(x) = \frac{1}{(2\pi)^{1/2}} \exp(-\frac{1}{2}x^2)$

Definition 2.8 (*Gradient of the discrete probability density function based on the Gaussian kernel*) Given a set of n observations of a variable x_l ($L = \{x_l\}_{l=1}^n, x_l \in \Re$) the discrete probability density function and it's gradient based on Gaussian Kernel are defined as $\hat{f}_G^L(x) = \frac{1}{nh(2\pi)^{1/2}} \sum_{l=1}^n \exp(-\frac{1}{2h^2}(x - x_l)^2)$

$\nabla \hat{f}_G^D(x) = \frac{\partial}{\partial x} \hat{f}_G^D(x) = \frac{1}{nh^3(2\pi)^{1/2}} \sum_{i=1}^n \exp[-\frac{1}{2h^2}(x - x_i)^2](x_i - x)$

The use of the Gaussian kernel allows to compute the maxima of the PDF based on a hill-climbing algorithm guided by the gradient of the PDF [12]. The bandwidth or smoothness parameter h, however, must be adjusted to compute the gradient.

The parameter h determines the influence of observations closer to one for which the PDF is evaluated, so it has an impact in the shape of the estimated PDF. The three mayor types of strategies to select a static bandwidth are: rule-of-thumb (ROT), cross-validation (CV) and plug-in (PI) selectors. The use of these selection approaches does not work well when the distribution of the variable is multimodal as it may be the case

for multimode processes. Therefore, an alternative approach to automatically selecting the bandwidth is to consider an adaptive approach, where two main approaches
can be used: balloon estimators, and sample point estimators [8].

Balloon estimators for the bandwidth present several practical application drawbacks being the most important one the fact that the estimated PDF does not integrate
to one. The sample point estimators are then selected in this book. The discrete PDF
as well as its gradient are thus redefined as follows

Definition 2.9 (*Probability density function discrete estimate with sample point estimators*) Given a set of n observations of a variable x_l ($L = \{x_l\}_{l=1}^n$, $x_l \in \Re$) the
estimate of the discrete probability density function is defined as

$$\hat{fsp}^L(x, \mathbf{h}) = \frac{1}{n} \sum_{l=1}^n \frac{1}{h(l)} K(\frac{1}{h(l)}(x - x_l)).$$

$$\nabla \hat{fsp}_G^L(x, \mathbf{h}) = \frac{\partial}{\partial x} \hat{fsp}_G^L(x, , \mathbf{h}) = \frac{1}{n(2\pi)^{1/2}} \sum_{l=1}^n \frac{1}{h(l)^3} \exp[-\frac{1}{2h(l)^2}(x - x_l)^2](x_l - x)$$

where K is a Kernel function and $\mathbf{h} \in \Re^n$ is a vector that contains the smoothness
parameter corresponding to each observation.

Algorithm 2.1 describes how to estimate the vector $\mathbf{h}$ in an automatic way. The
Matlab© implementation of this algorithm is shown in Appendix A.1. The first step
of the algorithm consists of estimating a fixed bandwidth based on the ROT [24].
The adaptive bandwidth for each observation is then estimated based on the fixed
bandwidth. The only parameter to be set is c. If c is set to 0 then the adaptive
estimation approach reduces to the fixed one. In practice, c can be set to 0.5 [8].

Algorithm 2.1 Algorithm to estimate the bandwidth vector for a variable

---

{Inputs: D: data set, $c \geq 0$: adaptation parameter}
{Output: $\mathbf{h}$: bandwidth vector}
$\mathbf{h} = \emptyset$
$IQR = Q_3 - Q_1$ {Calculate the interquartile range}
$\hat{\mu} = \frac{1}{n} \sum_{i=1}^n x_i$ {Estimate the mean}
$\hat{\sigma} = \frac{1}{n} \sum_{i=1}^n (x_i - \mu)^2$ {Estimate the variance}
$h = 1.06 \min\{\hat{\sigma}, \frac{IQR}{1.34}\} n^{-\frac{1}{5}}$ {Set an initial fixed bandwidth}
{Calculate the geometric mean of the PDF given a fixed bandwidth}
$T = \exp\left(\frac{1}{n} \sum_{i=1}^n \ln(\hat{f}_G^D(x_i, h))\right)$
{Evaluate the pdf according to the fixed bandwidth}
for all $x_i \in D$ **do**

$\quad \mathbf{h} = \mathbf{h} \bigcup h \left(\frac{\hat{f}_G^D(x_i, h)}{T}\right)^{-c}$

end for
return h

Once the bandwidth vector $\mathbf{h}$ has been estimated, the number of maxima can
be computed according to the algorithm presented in Algorithm 2.2. The Matlab©
implementation of this algorithm is shown in Appendix A.2.

There are two remaining parameters to calibrate in the algorithm: ε: convergence
threshold, and *maxit*: maximum number of iterations. ε and *maxit* determine the

---

Algorithm 2.2 Algorithm to find the number of maxima of the PDF of a data set

---

{Inputs: D: data set, $\mathbf{h}$: bandwidth parameter vector, ε: convergence threshold, $maxit$: maximum number of iterations}
{Output: $|\hat{X}^*|$: number of maxima}
$\hat{X}^* = \emptyset$ {Set of maxima}
for all $x_i \in D$ **do**
   $x^* = \mathbf{FindMaximum}(x_i, \mathbf{h}, D, \varepsilon, maxit)$
   if $x^* \notin \hat{X}^*$ **then**
      $\hat{X}^* = \hat{X}^* \bigcup x^*$
   end if
end for
return $|\hat{X}^*|$

$\mathbf{FindMaximum}(x_i, \mathbf{h}, D, \varepsilon, maxit)$
Initialization of iteration count: $it = 0$
$x^{(it)} = x_i$
repeat

$$x^{(it+1)} = \frac{\sum_{i=1}^{n} K_G(\frac{x^{(it)}-x_i}{h(i)})x_i}{\sum_{i=1}^{n} K_G(\frac{x^{(it)}-x_i}{h(i)})}$$

   $it = it + 1$
until $\| x^{(it)} - x^{(it-1)} \|^2 \le \varepsilon$ **or** $it == maxit$
return $x_i^{(it)}$

---

convergence of the algorithm, and they can be adjusted by examining the evolution of the term $\| x^{(it)} - x^{(it-1)} \|^2$ where it represents the iteration number. However, ε can be empirically set to a small value in the order of 10^{-4}, and $maxit$ can be set to 100–500 iterations depending on the computation capability available.

The two algorithms presented allow to determine in an automatic way if a variable has a multimodal distribution or not. Therefore, according to the previously presented definitions, if a single variable has a multimodal distribution, then the process can be considered as multimode from a data-driven perspective.

In the next sections, three benchmark case studies will be presented as examples of multimode processes.

2.3 Continuous Stirred Tank Heater (CSTH)

The Continuous Stirred Tank Heater (CSTH) process is a non-linear dynamic process whose mathematical model was created based on real data acquired from different sensors. The configuration of the CSTH process is shown in Fig. 2.4. The pilot plant used to create the model is located in the Department of Chemical and Materials Engineering at the University of Alberta [27]. Hot and cold water are mixed in the stirred tank experimental rig, and the water is heated by using steam that goes through a heating coil. The water is drained from the tank through a pipe. The temperature in the tank is assumed to be the same as the outflow temperature because the CSTH is well mixed.

Fig. 2.4 Continuous Stirred
Tank Heater technological
diagram from Thornhill et
al. [27]

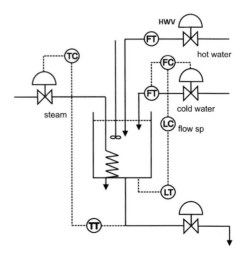

The dynamic volumetric and heat balance equations for this process are [27]

$$\frac{\partial}{\partial t}V(L) = f_{cw} + f_{hw} - f_{out}(L) \qquad (2.2)$$

$$\frac{\partial}{\partial t}H = W_{st} + h_{hw}\rho_{hw}f_{hw} + h_{cw}\rho_{cw}f_{cw} - h_{out}\rho_{out}f_{out}(L) \qquad (2.3)$$

where L represents the level, V the volume of water, f_{hw} the hot water flow into
the tank, f_{cw} the cold water flow into the tank, f_{out} the outflow from the tank, H
represents the total enthalpy in the tank, h_{hw} the specific enthalpy of hot water feed,
h_{cw} the specific enthalpy of cold water feed, h_{out} the specific enthalpy of the water
leaving the tank, ρ_{hw} the density of the incoming hot water, ρ_{cw} the density of the
incoming cold water, ρ_{out} the density of the water leaving the tank, and W_{st} the heat
income from steam.

The flow-meter installed in the cold water input pipe (FT) is an orifice plate with
differential pressure transmitters. The range for the flow-meter goes from 0 to 0.21
in litters per second (lps). The level sensor (LT)is also based on differential pressure
measurement with a range for the level of the tank from 0 to 50 cm. Finally, the
temperature sensor (TT) is a type J metal sheathed thermocouple with a range from
25 to 65 °C. All transmitters have a nominal 4–20 mA output, with a sampling rate
of 1 s. Therefore, all variables will be presented in mA.

The control scheme for this process is decentralized given its simplicity. All
controllers (Temperature (TC), Level (LC), and cold water input flow (FC)) are
standard Proportional-Integral controllers.

The simulations of the CSTH are hybrid in the sense that real noise sequences
are added to the outputs of the model [27]. The multimode operation of the CSTH
depends on the set-point of the controllers and the status of the hot water valve
(HWV). These features have motivated the use of the CSTH as a case study process
for data-driven fault diagnosis of multimode processes [18, 32].

Table 2.1 Operating modes of the CSTH

Variable (mA)	Mode 1	Mode 2	Mode 3	Mode 4	Mode 5
Level	12	12	12	13	13
Temperature	10.5	10.5	11	10.5	10.5
HWV status	0	5.5	5.5	5.5	5

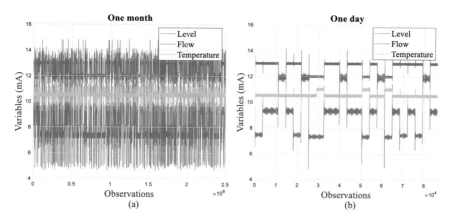

Fig. 2.5   The behavior of the three measured signals for one month (**a**) and one day (**b**)

2.3.1   Nominal Operation

Five operating modes are considered in this book, and they are presented in Table 2.1. One month of nominal data will be simulated, thus generating around 2.5 million of observations. A random mode change is simulated with a 50% probability, and a periodicity of one hour, to represent the natural fluctuations which may occur in a real system.

The measured signals are shown in Fig. 2.5. Observe that even when the multi-mode operation depends on the status of the hot water valve (HWV) this variable is not measured. This illustrates a problem of many systems where the variables that cause the multimode behavior are not measured. Besides the full set of 2.5 million observations (Fig. 2.5a), it is also shown a set for one day of observations, randomly chosen, just to show closely the variation of the variables (see Fig. 2.5b). If the previously presented algorithm to assess if the process is multimode will be applied on each variable of the entire data set, it would be extremely time consuming. However, if the algorithm is run on a set of continuous observations, and more than one maximum is detected, then the sufficient condition for multimode would be satisfied. The discrete PDF of each variable is shown in Fig. 2.6. When the algorithm to calculate the number of maxima is used for each variable, it is confirmed that this process is multimode because there is more than one maximum.

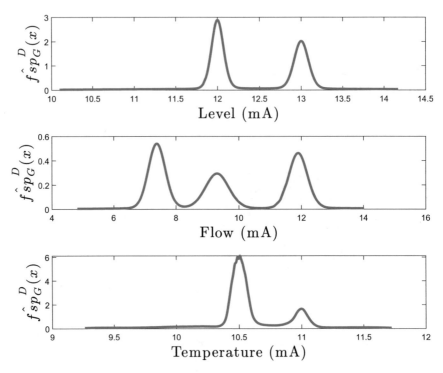

Fig. 2.6 Discrete PDF of each variable of CSTH process calculated with the sample-point estimator and the Gaussian kernel

2.3.2 Faulty Operation

A common fault that may arise in many industries is sensor calibration problems. This type of sensor fault is usually reflected in the measurements as a small bias in the signal of the variable. Such fault can be hard to detect depending on the normal range of operation of the variable. A data set comprising two months of operation is generated with five single abrupt sensor faults occurring in different moments, and with a different duration according to Table 2.2. Therefore, faults in different sensors are introduced with different duration. Each duration represents the time involved in detecting, diagnosing and recovering from the fault. This means that when a fault occurs it must be detected, but as soon as the problem is fixed the false alarms should be avoided. The changes of mode occur in the same way as in the previously generated data set. Therefore, it is not possible to know in advance in which mode the fault starts, and if a change of mode occurs while the fault is present.

Table 2.2 Faults of the CSTH

No	Sensor	Bias (mA)	Start (day)	Duration (days)
1	Level	0.5	3	3
2	Temperature	0.5	8	4
3	Level	0.5	15	2.3
4	Temperature	0.5	20	1
5	Level	0.5	23	3.5

Fig. 2.7 Hanoi water distribution network

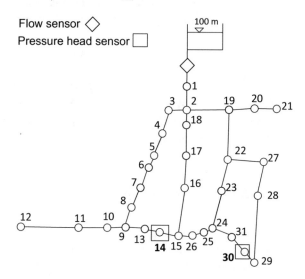

2.4 Hanoi Water Distribution Network (HWDN)

The water industry has the fundamental role in modern society of guaranteeing safe drinking water for the population. Water distribution networks (WDNs) are operated to guarantee drinking water to different types of consumers [9]. WDNs are dynamic non-linear systems formed by different elements: reservoirs, pipe networks, and water control elements. The variation of the variables in a WDN is determined by first-principle physical laws, the system layout, and consumers' demand.

The case study considered represents the planned water distribution trunk network of Hanoi (Vietnam). The configuration is shown in Fig. 2.7. The network is gravity-fed by a single reservoir, and it is formed by 34 pipes and 31 nodes. The pipe diameters are set according to Sedki and Ouazar, 2012 [23]. The link lengths and other parameters can be found in Ref. [7].

Flow in the pipes (q_i) and pressure head at the junctions (p_n) are physical variables which allow to characterize WDNs. Conservation of mass and energy are the main physical laws that describe the interactions among the variables. Flow is usually only measured in the input pipe of the system because pressure head sensors are cheaper and their installation and maintenance are easier to perform [13]. Given the size of the network only two pressure head sensors are considered. The positions of these sensors are nodes 14 and 30, because previous works have found these as good locations for fault diagnosis tasks [25]. Since there is a high cost with transmitting the signals at a high frequency, the sampling time considered in this book is 30 min.

The multimode behavior of the gravity-fed networks is dictated by the consumer's demand because these are the inputs of the system. The two main factors related to the consumer's demand are: the hour to hour variations, and the consumption patterns, which may vary from day to day. In general, the consumer's demand d_i at node i can be modeled as a stochastic variable formed by three components [2, 6]

$$d_i = d_i^{(s)} + d_i^{(p)} + d_i^{(c)} \tag{2.4}$$

where $d_i^{(s)}$ represents a seasonal component that resumes the long-term average daily water consumption, $d_i^{(p)}$ is a periodic component that represents the daily consumption pattern, and $d_i^{(c)}$ is a stochastic signal that represents the uncertain behavior of the consumers.

2.4.1 Nominal Operation

All simulations considered in this book for the HWDN benchmark problem are performed with the package EPANET 2.0 [22] and MATLAB 2018a©. To consider the realistic simulations of the network, the following conditions are considered:

1. Measurements uncertainty. Pressure head measurements are corrupted by additive noise with an amplitude within the range $[-0.25m, 0.25m]$.
2. Demand variability. The nominal conditions of the demand will change depending on the above-mentioned three components, represented in Eq. (2.4). Figure 2.8 shows an example for the three components of the demand. The first component is the seasonal one, and it is simulated as a monotonic linear slope. The average consumption increase represented in this component captures the seasonal consumption variability, e.g. Spring to Summer. The second component is the periodic consumption pattern that varies depending on the day of the week. Three daily patterns are simulated: one for weekdays, one for Saturday, and one for Sunday. Finally, the third component emulates the demand variability that cannot be explained by the other two components. Therefore, it is simulated as a stochastic variable with a Normal distribution $\mathcal{N} \sim \{0, \sigma\}$. Here, the standard deviation σ is related to the hour of the day because the higher uncertainty is related to the maximum consumption periods. This relationship is empirically generated by considering the relationship $\sigma(t) = 0.025 \star d_i^{(p)}(t)$. The uncertainty will be higher when the periodic component consumption is also high.

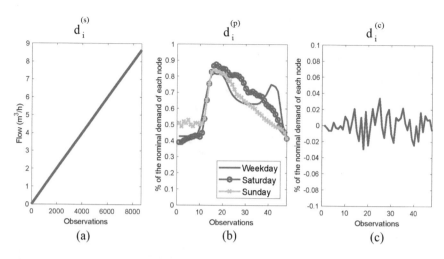

Fig. 2.8 Components of the consumer's demand. **a** Seasonal component (six months), **b** Periodic component: three different daily consumtion patterns, **c** Stochastic component (one day)

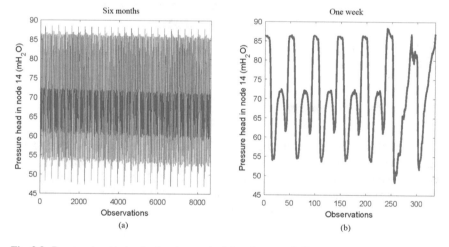

Fig. 2.9 Pressure head behavior for six months (**a**) and one week (**b**)

A data set comprising six months with approximately 8640 observations for each monitored variable (given the sampling time) is considered. The features of the demand generate measured signals like the ones shown in Fig. 2.9. Besides the full set of 8640 observations for the pressure head at the node 14, Fig. 2.9a, it is also shown a set for one week of observation, randomly chosen, for the same monitored variable, just to provide a close view of its variation (see Fig. 2.9b). The estimated discrete PDF of each variable is shown in Fig. 2.6. Once again, when the algorithm to calculate the number of maximums is used for each variable it is confirmed that this process is multimode because there is more than one maximum (Fig. 2.10).

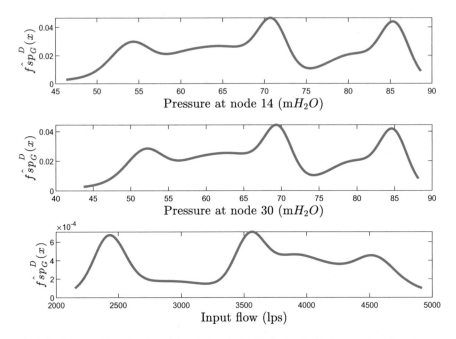

Fig. 2.10 Discrete PDF of each variable of Hanoi WDN calculated with the sample-point estimator and the Gaussian kernel

2.4.2 Faulty Operation

The most important faults to be considered for WDNs are leakages. Although leakages can occur in any pipe of the network, it is considered in most works, for simplification purposes, that they occur in the nodes. If a leak occurs in a pipe, it affects the mass-balance equation in the nodes. Thus, the leakage outflow caused by a leak that occurs in a node f_i has an effect in Eq. (2.4) as follows

$$\sum_{j=1}^{b_{ij}} q_{ij} = d_i + f_i \tag{2.5}$$

where d_i is the consumer's demand and b_{ji} is the number of branches connected to the node i. In the derivation of Eq. (2.4), i.e. the mass balance at node i, are considered that all flows are incoming to the node, except the demand and the leak. It is clear from the previous equation that the main task of fault diagnosis techniques is to distinguish the changes caused by the consumer's demand from leakages [9]. There is also a non-linear relationship between pressure head and outflow caused by leakages described by the following equation

$$f_i = C_e h_n^{\gamma} \tag{2.6}$$

Table 2.3 Emitter coefficient for each data set

	Data set					
	1	2	3	4	5	6
C_e	10	12	16	20	23	28

where the emitter coefficient C_e is associated with the pipe rupture size and $\gamma = 0.5$ [22]. Therefore, the characterization of pressure head behavior measured at different nodes of the network, together with the input flow, allows to diagnose this type of fault.

A total of six data sets with six months of data in each one will be considered for testing the fault diagnosis performance. A total of 30 leakage scenarios (one for each node, with the exception of the tank node) are generated for each data set and all the leakage scenarios will have an emitter coefficient value according to Table 2.3, with a random duration between one and seven days. The leak amount depends on the pressure head and the emitter coefficient value, that varies between 10 and 28. Consequently, the leakage outflow simulated (25–75 lps) varies between (0.45% and 1.39% of the maximum water demanded (5,539 lps).

2.5 Tennessee Eastman Process (TEP)

The case study known as Tennessee Eastman process (TEP) represents a non-linear dynamic chemical process of an industry, and it was created by the Eastman Chemical Company with the goal of evaluating the performance of control and fault diagnosis techniques [5]. Figure 2.11 shows the configuration of the process. The TEP plant is formed by five operating units: a reactor, a product condenser, a recycle compressor, a vapor-liquid separator and a product stripper. The plant is designed to manufacture two products G and H from four reactants A, C, D and E. The byproduct F also results from these reactions. The chemical reactions that describe the process are

$$A_{(g)} + C_{(g)} + D_{(g)} \rightarrow G_{(Liq)} \quad Product\ 1 \tag{2.7}$$

$$A_{(g)} + C_{(g)} + E_{(g)} \rightarrow H_{(Liq)} \quad Product\ 2 \tag{2.8}$$

$$A_{(g)} + E_{(g)} \rightarrow F_{(Liq)} \quad Byproduct \tag{2.9}$$

$$3D_{(g)} \rightarrow 2F_{(Liq)} \quad Byproduct \tag{2.10}$$

A decentralized control strategy is used because less variability in the production volume and product quality is guaranteed under nominal conditions for different operating points [16]. The number of variables measured is 53. From this total there are 19 quality variables measured, 14 of them with a sampling rate of 0.1 h, and 5 of them with a sampling rate of 0.25 h. The rest of the variables are measured with a sampling rate of 0.01 h. A total of 31 variables out of the 34 which are measured

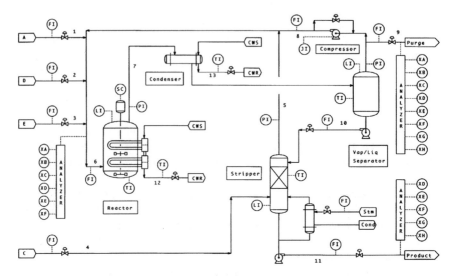

Fig. 2.11 Tennessee Eastman process technological diagram from Downs and Vogel, 1993 [5]

at a fast sampling rate will be used for fault diagnosis. The three excluded variables are the recycle valve, steam valve and agitator because their behavior is completely independent of the faults and operating modes that characterize the TEP [38, 39].

2.5.1 Nominal Operation

The TEP is designed to work under the six operating modes shown in Table 2.4. The change of operating mode depends on the desired properties of the product: G/H mass ratio and production volume (kg/h) [21]. When a mode change is activated the set-point of multiple controllers change accordingly. Given that the TEP presents a non-linear, dynamic and multimode behaviour, it has been widely used for testing the performance of fault diagnosis techniques [4, 18, 28, 30, 38]. In the simulations, Gaussian noise is added to all process measurements with zero mean and standard deviation typical of the measurement type.

Modes 1, 2 and 3 are selected for continuous simulations. The process runs initially in mode 3 for 10 h, and changes to mode 1. After 60 h, mode 2 is activated for 70 h and finally mode 3 is activated again for the last 100 h. Transitions in this process may take several hours due to safety constraints mainly associated with avoiding large pressure variations in the reactor. The last transition takes around 40 h to be completed because of the drastic change in the desired characteristics of the final product.

A data set of 200 h of process operation is generated with a total of observations around 20,000 (sampling time of 0.01 h). The transition between two steady modes

Table 2.4 Operating modes of the TEP

Mode	G/H mass ratio	Production volume (kg/h)
1	50/50	14076
2	10/90	14077
3	90/10	11111
4	50/50	Maximum
5	10/90	Maximum
6	90/10	Maximum

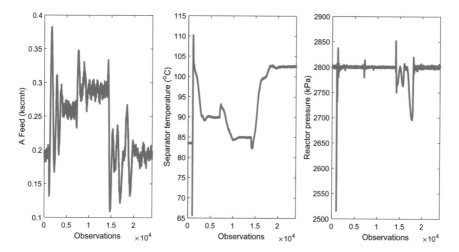

Fig. 2.12 Behavior of three variables for 200 h of process operation

lasts for hours because of the slow dynamics of the chemical reaction. The 31 measured variables have different units and they present a different behavior that does not allow to manually separate the steady modes from the transitions by visual inspection. This fact can be confirmed in Fig. 2.12 where the behavior of three variables is shown.

The estimated discrete PDF of three variables is shown in Fig. 2.13. It is then confirmed for this process that it is multimode because there is more than one maximum. It must be remarked that it is very difficult to determine the number of modes because the number of variables is high. Strategies to determine the number of modes are proposed in the next chapter.

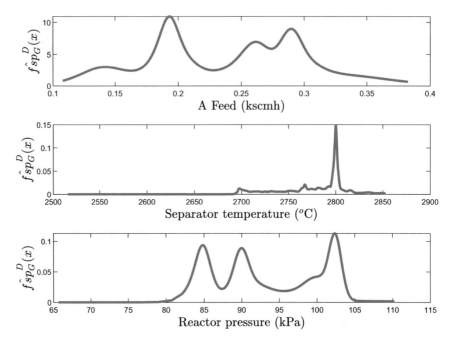

Fig. 2.13 Discrete PDF of three variables of TEP calculated with the sample-point estimator and the Gaussian kernel

2.5.2 Faulty Operation

A set of 21 faults can be considered for the TEP. These are caused by abrupt changes or random variability of some variables, sticking valves, slow drift in the reaction, and unknown causes. Six faults scenarios will be considered. Table 2.5 presents each of these scenarios. The numbers in parenthesis in the table correspond to the fault numbers. These six faults will be simulated in steady modes 3 and 1, as well as during the transition between these two modes. Each fault will be simulated for 500 observations after they occur representing around 5 h of operation.

Remarks

In this chapter, the definition of multimode process was presented from a data-driven standpoint. An algorithm to determine in an automatic fashion if a process is multimode is also presented. The three cases of study which will be used through the book are described. The nominal and fault operation of each process is discussed. Finally, it is demonstrated that each process is multimode by applying the presented algorithm.

Table 2.5 Simulation fault scenarios for the TEP

Fault scenario	Description	Type
1 (1)	A/C feed ratio, B composition constant (stream 4)	Step
2 (5)	Condenser water inlet temperature	Step
3 (10)	C feed temperature (stream 4)	Random variation
4 (13)	Reaction kinetics	Slow drift
5 (14)	Reactor cooling water valve	Sticking
6 (16)	Unknown	Unknown

References

1. AlGhazzawi, A., Lennox, B.: Monitoring a complex refining process using multivariate statistics. Control Eng. Pract. **16**, 294–307 (2008)
2. Alvisi, S., Franchini, M., Marinelli, A.: A short-term, pattern-based model for water-demand forecasting. J. Hydroinform. **9**(1), 39–50 (2007). https://doi.org/10.2166/hydro.2006.016.. https://iwaponline.com/jh/article/9/1/39/31287/A-shortterm-patternbased-model-for-waterdemand
3. Cao, S., Rhinehart, R.R.: An efficient method for on-line identification of steady-state. J. Process Control **5**(6), 363–374 (1995)
4. Chiang, L.H., Russell, E.L., Braatz, R.D.: Fault Detection and Diagnosis in Industrial Systems. Springer (2001)
5. Downs, J.J., Vogel, E.F.: A plant-wide industrial problem process. Comput. Chem. Eng. **17**(3), 245–255 (1993)
6. Eliades, D.G., Polycarpou, M.M.: Leakage fault detection in district metered areas of water distribution systems. J. Hydroinform. **14**(4), 992–1005 (2012)
7. Fujiwara, O., Khang, D.B.: A two-phase decomposition method for optimal design of looped water distribution networks. Water Resour. Res. **26**(4), 539–549 (1990)
8. Gramacki, A.: Nonparametric Kernel Density Estimation and Its Computational Aspects. Springer (2018)
9. Quiñones Grueiro, M., Verde, C., Llanes-Santiago, O.: Features of Demand Patterns for Leak Detection in Water Distribution Networks. Chapter 9, pp. 171–189. Springer, Cham, Switzerland (2017)
10. Ha, D., Ahmed, U., Pyun, H., Lee, C.j., Baek, K.H., Han, C.: Multi-mode operation of principal component analysis with k-nearest neighbor algorithm to monitor compressors for liquefied natural gas mixed refrigerant processes. Comput. Chem. Eng. **106**, 96–105 (2017)
11. He, Y., Ge, Z., Song, Z.: Adaptive monitoring for transition process using dynamic mutual information similarity analysis. In: Chinese Control and Decision Conference, pp. 5832–5837 (2016)
12. Hinneburg, A., Keim, D.A.: A general approach to clustering in large databases with noise. Knowl. Inform. Syst. **5**(4), 387–415 (2003)
13. Jung, D., Lansey, K.: Water distribution system burst detection using a nonlinear kalman filter. J. Water Resour. Plann. Manag. **141**(5), 1–13 (2015). https://doi.org/10.1061/(ASCE)WR.1943-5452.0000464
14. Kelly, J.D., Hedengren, J.D.: A steady-state detection (SSD) algorithm to detect non-stationary drifts in processes. J. Process Control **23**(3), 326–331 (2013)

15. Kodamana, H., Raveendran, R., Huang, B.: Mixtures of probabilistic PCA with common structure latent bases for process monitoring. IEEE Trans. Control Syst. Technol. (2017). https://doi.org/10.1109/TCST.2017.2778691
16. Lawrence Ricker, N.: Decentralized control of the Tennessee Eastman challenge process. J. Process Control **6**(4), 205–221 (1996)
17. Maestri, M., Farall, A., Groisman, P., Cassanello, M., Horowitz, G.: A robust clustering method for detection of abnormal situations in a process with multiple steady-state operation modes. Comput. Chem. Eng. **34**(2), 223–231 (2010)
18. Ning, C., Chen, M., Zhou, D.: Hidden Markov model-based statistics pattern analysis for multimode process monitoring: an index-switching scheme. Ind. Eng. Chem. Res. **53**(27), 11084–11095 (2014)
19. Olsson, G.: Instrumentation, control and automation in the water industry -state-of-the-art and new challenges. Water Sci. Technol. **53**(4), 1–16 (2006)
20. Papoulis, A.: Probability, Random Variables and Stochastic Processes (1991)
21. Ricker, N.L.: Optimal steady-state operation of the Tennessee Eastman challenge process. Comput. Chem. Eng. **19**(9), 949–959 (1995)
22. Rossman, L.A.: Water supply and water resources division. National Risk Management Research Laboratory. Epanet 2 User's Manual. Technical report United States Environmental Protection Agency (2000). http://www.epa.gov/nrmrl/wswrd/dw/epanet.html
23. Sedki, A., Ouazar, D.: Hybrid particle swarm optimization and differential evolution for optimal design of water distribution systems. Adv. Eng. Inform. **26**(3), 582–591 (2012)
24. Silverman, B.W.: Density Estimation for Statistics and Data Analysis. CRC Press (1986)
25. Soldevila, A., Fernandez-canti, R.M., Blesa, J., Tornil-sin, S., Puig, V.: Leak localization in water distribution networks using Bayesian classifiers. J. Process Control **55**, 1–9 (2017)
26. Srinivasan, R., Wang, C., Ho, W.K., Lim, K.W.: Dynamic principal component analysis based methodology for clustering process states in Agile chemical plants. Ind. Eng. Chem. Res. **43**(9), 2123–2139 (2004)
27. Thornhill, N.F., Patwardhan, S.C., Shah, S.L.: A continuous stirred tank heater simulation model with applications. J. Process Control **18**, 347–360 (2008)
28. Tong, C., Yan, X.: Double monitoring of common and specific features for multimode process. As. Pac. J. Chem. Eng. **8**(5), 730–741 (2013)
29. Vazquez, L., Blanco, J.M., Ramis, R., Peña, F., Diaz, D.: Robust methodology for steady state measurements estimation based framework for a reliable long term thermal power plant operation performance monitoring. Energy **93**, 923–944 (2015)
30. Wang, F., Tan, S., Peng, J., Chang, Y.: Process monitoring based on mode identification for multi-mode process with transitions. Chemom. Intell. Lab. Syst. **110**(1), 144–155 (2012)
31. Wang, X., Wang, X., Wang, Z., Qian, F.: A novel method for detecting processes with multi-state modes. Control Eng. Pract. **21**, 1788–1794 (2013)
32. Xu, X., Xie, L., Wang, S.: Multimode process monitoring with PCA mixture model. Comput. Electr. Eng. **40**(7), 2101–2112 (2014)
33. Xu, Y., Deng, X.: Fault detection of multimode non-Gaussian dynamic process using dynamic Bayesian independent component analysis. Neurocomputing **200**, 70–79 (2016)
34. Yu, H.: A novel semiparametric hidden Markov model for process failure mode identification. IEEE Trans. Autom. Sci. Eng. **15**(2), 506–518 (2017)
35. Yu, J.: A nonlinear kernel Gaussian mixture model based inferential monitoring approach for fault detection and diagnosis of chemical processes. Chem. Eng. Sci. **68**(1), 506–519 (2012)
36. Yu, J., Qin, S.J.: Multimode process monitoring with Bayesian inference-based finite Gaussian mixture models. AIChE J. **54**(7), 1811–1829 (2008)
37. Zhao, S.J., Zhang, J., Xu, Y.M.: Performance monitoring of processes with multiple operating modes through multiple PLS models. J. Process Control **16**, 763–772 (2006)
38. Zhu, Z., Song, Z., Palazoglu, A.: Transition process modeling and monitoring based on dynamic ensemble clustering and multiclass support vector data description. Ind. Eng. Chem. Res. **50**(24), 13969–13983 (2011)
39. Zhu, Z., Song, Z., Palazoglu, A.: Process pattern construction and multi-mode monitoring. J. Process Control **22**, 247–262 (2012)

Chapter 3
Clustering for Multimode Continuous Processes

Abstract This chapter presents the use of clustering methods for labeling the data of multimode processes. Fault diagnosis represents a twofold challenge for multimode continuous processes: the unlabeled data must be characterized with respect to the number of modes (which form clusters or groups), and diagnosis schemes must be developed by using the labeled data. The former task is challenging because of the uncertainty related to the unknown number and types of modes. Therefore, the advantages and drawbacks of clustering solutions applied to multimode processes are discussed from the theoretical and practical point of view.

3.1 Clustering Analysis for Multimode Processes

Once a process has been identified as multimode according to the method presented in the previous chapter, the following characteristics must be identified to be able to design the fault detection strategies:

- Number of modes.
- Mode type.
- Observations corresponding to each mode.

The goal of clustering methods is to separate a data set into different groups such that the observations are organized according to a similarity criteria [12, 26]. Formally, the clustering task for continuous variables can be defined as follows:

Definition 3.1 (*Clustering task for continuous variables*) Given a data set formed by n_n observations in a m dimensional space ($D_n = \{\mathbf{x}\}_{j=1}^{n_n}$, $\mathbf{x}_j \in \Re^m$) that is representative of the normal/typical operation of a process find the set of clusters $C = \{C_1, C_2, ..., C_k\}$ such that

- $P(\mathbf{x}_j \in C_i) = \varrho_{ji}$
- $\sum_{i=1}^{k} \varrho_{ji} = 1$

where ϱ_{ji} represents the probability of observation $\mathbf{x}_j$ belong to cluster C_i.

© Springer Nature Switzerland AG 2021   35
M. Quiñones-Grueiro et al., *Monitoring Multimode Continuous Processes*,
Studies in Systems, Decision and Control 309,
https://doi.org/10.1007/978-3-030-54738-7_3

When hard clustering is performed, it means that an observation can only belong to one group, such that $\varrho_{ji} = 1$ for cluster i and $\varrho_{ji} = 0$ for any other group or cluster in C. On the other hand, soft clustering considers that an observation has a different probability of belonging to each group.

Clustering methods are considered unsupervised learning tools because the label of the observations is unknown in advance. Therefore, clustering analysis is used to identify the features of multimode processes. The following procedure can be followed to apply clustering methods to data from multimode processes:

1. Select the features from the data that will be used as input for the clustering method. Different feature selection techniques can be used as a previous step to clustering methods [1, 6]. The goal of this step is to select the set of features which may contain relevant information regarding the true structure of the data. Thus, these features will allow to recover such structure. For example, entropy-based similarity measures can be used for feature selection based on the notion that two observations belonging to the same cluster or different clusters, will contribute to the total entropy less than if they were uniformly separated. In many fields, expert knowledge is available such that the features selection is tailored to the specific problem.

2. Preprocess the data. The goal of data preprocessing is to remove the characteristics of the data set that may lead to a biased clustering result. The three main tasks to consider are:

   - Outlier removal. This step is required whenever the data may contain outliers because many clustering algorithms are not robust against outliers. This means that some outliers may cause undesired clustering results.
   - Noise filtering. This step is required if the significant noise affects the signals. It is specially important when using clustering methods that are sensible to varying densities in the data.
   - Data scaling. This step is generally required when the units of the variables/signals are different. The goal is to give all variables an equal importance when applying the clustering method.

3. Select a set of evaluation metrics for deciding the parameters of the clustering method. The evaluation metrics are tools used to decide the quality of the clustering results. Therefore, they are required in the selection of the clustering parameters as well as to decide the final structure identified in the data set.

4. Select and apply a clustering method. There are many possible clustering methods to be used for multimode data. In Sect. 3.4, the most relevant clustering methods used for multimode processes are discussed in detail.

5. Analyze the results through clustering evaluation metrics which allow to judge the quality of the clustering results. Therefore, if the performance is not satisfactory, go back to step 1 and modify either the features to be analyzed, the clustering method, or the clustering method parameters.

3.2 Preprocessing Data for Clustering Multimode Data

The most important preprocessing step for multimode processes is data scaling. Therefore, the strategies to scale the data are discussed below. No feature selection method or data filtering technique is used in this book to cluster multimode data. The reader is, however, referred to Ref. [41] to find further information about outlier removal and data filtering methods.

It is generally recommended to scale the variables to be clustered as a prepro-cessing step [12]. The main idea behind scaling is to give all attributes an equal weight, thus reducing the influence of measurement units and/or magnitudes in the results. This facilitates recovering the underlying clustering structure. There are two common strategies to scale a variable:

- Min-Max:

$$x^* = \frac{x - min(x)}{max(x) - min(x)} \tag{3.1}$$

where $min(x)$ and $max(x)$ are the minimum and maximum values that characterize the variable probability distribution.
- Z-score:

$$x^* = \frac{x - \mu}{\sigma} \tag{3.2}$$

where μ and σ represent the mean and standard deviation estimated for the variable x, given the data set.

Previous studies have compared the effect of these two scaling strategies for clustering tasks [21]. Overall, Min-Max scaling seems to unanimously provide for the best recovery of the true cluster structure. Conversely, Z-score scaling was not especially effective. Therefore, Min-Max scaling is recommended for multimode data, and it will be used throughout the rest of the book unless explicitly noticed.

3.3 Clustering Evaluation Analysis

Evaluation metrics, also known as validity indexes, are used for the analysis of the clustering results. Each index evaluates the performance of the clustering results in a different way. Evaluation metrics allow to find the best parameters of the clustering methods to uncover the true structure of the data.

Clustering evaluation measures can be classified into [1, 21]

- **External indexes**: Use external information to evaluate whether the clustering structure resulting from an algorithm matches some external structure. Usually, the structure of the clusters discovered is evaluated based on some criteria either given by an expert or found in a reference labeled data set.

- **Internal indexes**: Use two primary criteria to evaluate the quality of a clustering result:

  – Compactness: Measures how closely related the observations in a cluster are.
  – Separability: Measures how well-separated a cluster is from other clusters.

  Given that the goal of clustering is to group similar observations within the same cluster, and separate different observations into distinct clusters, a similarity measure must be selected to define mathematically compactness and separability. For instance, the Euclidean distance between two observations could be used for this purpose.

Since a reference labeled data set is generally not available for multimode processes, internal indexes are generally used. The premise behind all internal indexes is to evaluate the separability and compactness of the resulting clusters based on the concept of inter-cluster and intra-cluster distance, respectively. These distances can be defined as follows

Definition 3.2 (*Average intra-cluster distance*) Given a set of clusters $C = \{C_1, C_2, ..., C_k\}$, the average intra-cluster distance is

$$\delta^{intra} = \frac{1}{k} \sum_{i=1}^{k} \delta_i^{intra} \tag{3.3}$$

where δ_i^{intra} is defined for each cluster C_i as

$$\delta_i^{intra} = \frac{1}{|C_i|(|C_i| - 1)} \sum_{\substack{\mathbf{x}_l, \mathbf{x}_m \in C_i \\ l \neq m}} \delta(\mathbf{x}_l, \mathbf{x}_m) \tag{3.4}$$

δ is a similarity measure such as the Euclidean distance, $\mathbf{x}_l$ and $\mathbf{x}_m$ represent two different observations belonging to cluster C_i, and $|C_i|$ represents de cardinality of the cluster C_i. The value of δ_i^{intra} gives an idea on how compact the cluster is. The cluster will be more compact when the value of δ_i^{intra} is smaller.

Other variant for the intra-cluster distance is the average of the distance between the two most remote objects belonging to the same cluster.

Definition 3.3 (*Average inter-cluster distance*) Given a set of clusters $C = \{C_1, C_2, ..., C_k\}$ the average inter-cluster distance is

$$\delta^{inter} = \frac{1}{k(k - 1)} \sum_{\substack{C_i, C_j \in C \\ i \neq j}} \delta_{ij}^{inter} \tag{3.5}$$

where δ_{ij}^{inter} is defined for two different clusters C_i and C_j as

Fig. 3.1 Three clustering results. **a** small intra-distance with small inter-distance, **b** large intra-distance with large inter-distance, and **c** small intra-distance with large inter-distance

$$\delta_{ij}^{inter} = \frac{1}{|C_i||C_j|} \sum_{\substack{\mathbf{x}_l \in C_i \\ \mathbf{x}_m \in C_j}} \delta(\mathbf{x}_l, \mathbf{x}_m) \tag{3.6}$$

δ is a similarity measure such as the Euclidean distance, and $\mathbf{x}_l$ and $\mathbf{x}_m$ represent two observations belonging to different clusters. The average inter-cluster distance provides a notion of how close/far the clusters are located.

Other variants for the inter-cluster distance are the average of the distance between two most remote objects belonging to two different clusters, or the closest distance between two objects belonging to two different clusters.

Figure 3.1 shows three possible clustering configurations. In Fig. 3.1a the clusters are compact, which is a desirable property, but their inter-distance is small meaning that the separability is not good. In Fig. 3.1b the clusters are well separated but the intra-cluster distance is large meaning that the clusters are not compact. Finally, in Fig. 3.1c the best result is shown where a small intra-cluster distance and a large inter-cluster distance are achieved.

Several evaluation metrics have been formulated based on these notions of separability. Previous studies have been developed to find the indexes that allow to evaluate the clustering methods independently of the features of the data. Three widely used internal indexes will be applied throughout this book, given the good results shown in previous works, and the availability in most mathematical and statistical software packages [3, 7, 20, 27, 30]: Silhouette coefficient, Calinski-Harabasz index and DaviesBouldin index.

- Silhouette coefficient:

$$Sil = \frac{1}{n_m} \sum_{i=1}^{n_m} s_i \quad , \quad s_i = \frac{b_i - a_i}{\max\{a_i, b_i\}} \tag{3.7}$$

with $\mathbf{x}_i$, $x_j \in C_m$ and

$$a_i = \frac{1}{n_m - 1} \sum_{\mathbf{x}_j \in C_m, i \neq j} \delta(\mathbf{x}_i, \mathbf{x}_j) \tag{3.8}$$

$$b_i = \min \frac{1}{n_l} \sum_{\mathbf{x}_j \in C_l} \delta(\mathbf{x}_i, \mathbf{x}_j) \text{ for } l = 1, 2, ..., k \text{ with } l \neq m \tag{3.9}$$

where n_m and n_l are the number of observations in clusters l and m, respectively, δ is a similarity measure (usually the Euclidean distance is used), a_i is the average similarity between $\mathbf{x}_i$ and the observations of its own cluster, and b_i is the smallest average distance to all observations in any other cluster. Overall, the smaller a_i the better the assignment. The closest s_i is to 1 the observation is appropriately clustered ($-1 \leq s_i \leq 1$). The same logic applies to Sil because it represents the average Silhouette value. The Matlab© implementation of this evaluation measure is shown in Appendix C.1.

- Calinski-Harabasz index:

$$CK = \frac{Trace(\mathbf{S_b})}{k-1} \bigg/ \frac{Trace(\mathbf{S_w})}{n-k} \tag{3.10}$$

with

$$\mathbf{S_b} = \sum_{i=1}^{k} n_i (\mathbf{c}_i - \mathbf{c_j})(\mathbf{c}_i - \mathbf{c_j})^T \tag{3.11}$$

$$\mathbf{S_w} = \sum_{i=1}^{k} \sum_{j=1}^{n} u_{ij} (\mathbf{x}_j - \mathbf{c}_i)(\mathbf{x}_j - \mathbf{c}_i)^T \tag{3.12}$$

where $u_{ij} \in [0, 1]$, $\mathbf{c}_i$ is the centroid of cluster i, $\mathbf{c}$ is the general mean of all centroids $\mathbf{c}_i$ and $\mathbf{S_b}$ and $\mathbf{S_w}$ are called between and within-class scatter matrix, respectively. The higher the value of CK the best partitioning of the data set because it is desirable that the between distance among different clusters' observations is higher than the within distance among observations of each cluster. The Matlab© implementation of this evaluation measure is shown in Appendix C.2.

- DaviesBouldin index:

$$DB = \frac{1}{k} \sum_{i=1}^{k} \max \left(\frac{\hat{d}_i + \hat{d}_j}{d(\mathbf{c}_i, \mathbf{c}_j)} \right) \text{ for } j = 1, 2, ..., k \text{ with } i \neq j \tag{3.13}$$

where c_i and c_j are the centroids of clusters i and j respectively, $\hat{d}_i$ represents the average Euclidean distance of all elements of cluster i to its centroid, and $d(\mathbf{c}_i, \mathbf{c}_j)$ is the Euclidean distance between the two centroids. The smallest Davies-Bouldin index indicates the best clustering configuration because it is desired to have small intra-cluster distances (high intra-cluster similarity or cluster homogeneity) and large inter-cluster distances (low inter-cluster similarity). The Matlab© implementation of this evaluation measure is shown in Appendix C.3.

3.4 Clustering Methods

Many clustering techniques have been applied for dividing the data of multimode processes into different operating modes. Figure 3.2 summarizes the four main clustering methods that have been applied for mode labeling of multimode processes [25]. The methods can be categorized into four branches: K-means, Fuzzy C-means, mixture modeling methods, and window-based approaches. Each approach will be discussed in more detail in the following subsections.

3.4.1 K-Means

Given n_n observations in a m dimensional space $(D_n = \{\mathbf{x}\}_{j=1}^{n_n}, \mathbf{x}_j \in \Re^m)$ and a number of groups k to cluster the data, the objective of the K-means clustering algorithm is to form theset of groups $C = \{C_1, C_2, ..., C_k\}$, such that the following optimization criterion is minimized [48]

$$J(k) = \sum_{i=1}^{k} \sum_{j=1}^{n_i} \delta(\mathbf{x}_j, \mathbf{c}_i) \tag{3.14}$$

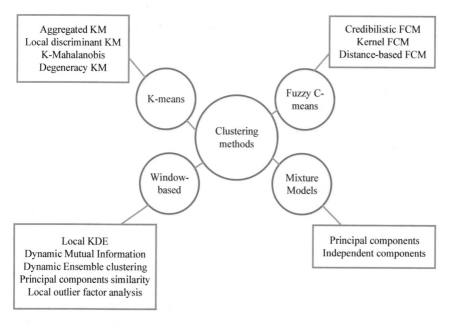

Fig. 3.2 Clustering methods used for mode labeling of multimode processes

where n_i is the number of observations of group i, $\delta(\mathbf{x}_j, \mathbf{c}_i)$ is a similarity measure, and $\mathbf{c}_i$ represents a vector that contains the expected value of each variable of group i. This vector is called centroid, and it is approximated by the mean of each cluster

$$\mathbf{c}_i = \frac{1}{n_i} \sum_{j=1}^{n_i} u_{ij} \mathbf{x}_j \tag{3.15}$$

where $u_{ij} \in [0, 1]$, $u_{ij} = 1$ if the observation $\mathbf{x}_j$ is a member of cluster C_i, otherwise $u_{ij} = 0$. As a result, a crisp partition of the space of observations is obtained.

Overall, the main parameters to be adjusted for the K-means method are the similarity metric and the number of clusters. The Euclidean (Eq. 3.16) and Mahalanobis (Eq. 3.17) distances are among the most popular similarity metrics used for this purpose.

$$\delta(\mathbf{x}_j, \mathbf{c}_i) = \parallel \mathbf{x}_j - \mathbf{c}_i \parallel^2 \tag{3.16}$$

$$\delta(\mathbf{x}_j, \mathbf{c}_i) = (\mathbf{x}_j - \mathbf{c}_i)^T \Sigma_i^{-1} (\mathbf{x}_j - \mathbf{c}_i) \tag{3.17}$$

where Σ_i^{-1} is the inverse covariance matrix estimated for cluster C_i. The K-means algorithm is presented in Algorithm 3.1. The Matlab$^{©}$ implementation of this algorithm is shown in Appendix D.1.

Algorithm 3.1 K-means clustering algorithm

{Inputs: D: data set, k: number of clusters, ε: convergence threshold, $maxit$: maximum number of iterations}
{ Initialization: $it = 0$, random initialization of k centroids }
repeat
    $it = it + 1$
    $C_i = \emptyset$ for $i = 1, 2, ..., k$
    {Cluster assignment step}
    for all $\mathbf{x}_j \in D$ **do**
        $j = \arg\min_i \{\delta(\mathbf{x}_j, \mathbf{c}_i)\}$ {Assign $\mathbf{x}_j$ to the closest centroid}
        $C_i = C_i \bigcup \{\mathbf{x_j}\}$
    end for
    {Centroid update step}
    for all $i = 1, 2, ..., k$ **do**
        $\mathbf{c}_i^{(it)} = \frac{1}{n_i} \sum_{j=1}^{n_i} u_{ij} \mathbf{x}_j$
    end for
until $\sum_{i=1}^{k} \parallel \mathbf{c}_i^{(it)} - \mathbf{c}_i^{(it-1)} \parallel^2 \leq \varepsilon$ **or** $it = maxit$

To identify the number of clusters (modes for multimode process data), multiple runs of the method are required together with a clustering validity criteria. Separability metrics among the clusters are used as validity criteria to determine the number of clusters: the Calinski-Harabasz [3], Davies-Bouldin [7], Silhouette index [27].

The main advantages of the K-means method are simplicity, computational efficiency, and easy implementation [15]. Therefore, it has been widely used for multimode clustering tasks. Some extensions of the traditional method are aggregated K-means, local discriminant regularized soft K-means, K-Mahalanobis and degeneracy on K-means [8, 23, 36, 37, 49]. The identification of clusters corresponding to steady and transition modes is the most challenging task for K-means-based methods. The main reason is that finding clusters with a non-convex shape or asymmetric size is a difficult task [42, 48]. Hence, only a few works tackle the identification of transition modes [37, 49].

3.4.2 Fuzzy C-Means

Given n_n observations in a m dimensional space $(D_n = \{\mathbf{x}\}_{j=1}^{n_n}, \mathbf{x}_j \in \mathfrak{R}^m)$ and a number of groups k to cluster the data, the objective of the Fuzzy C-means (FCM)clustering algorithm is to form the set of groups $C = \{C_1, C_2, ..., C_k\}$, such that the following optimization criterion is minimized

$$J(k, w) = \sum_{i=1}^{k} \sum_{j=1}^{n} u_{ij}{}^{\omega} d_{ij}(\mathbf{x}_j, \mathbf{c}_i)^2 \qquad (3.18)$$

$$\textbf{subject to } 0 < u_{ij} < 1 \textbf{ and } \sum_{i=1}^{k} u_{ij} = 1, \text{ for } \forall j \qquad (3.19)$$

where u_{ij} represents the membership degree of an observation $\mathbf{x}_j$ to the cluster C_i, such that the matrix $U \in \mathfrak{R}^{k \times n}$ denotes the fuzzy partition, and the exponent $\omega > 1$ regulates the fuzziness degree of the resulting partition. The closer the exponent ω is to 1, the closer to a crisp partition is the fuzzy partition. The higher the exponent, the more vague is the boundary among clusters. The distance d_{ij} can be calculated as follows

$$d_{ij}(\mathbf{x}_j, \mathbf{c}_i) = (\mathbf{x}_j - \mathbf{c}_i)^T A_i (\mathbf{x}_j - \mathbf{c}_i) \qquad (3.20)$$

where $\mathbf{c}_i$ represents the centroid of cluster C_i and $A_i \in \mathfrak{R}^{m \times m}$ can take different values depending on the distance metrics used (Euclidean: A_i is the identity matrix, Mahalanobis: A_i is the inverse covariance matrix calculated with respect to $\mathbf{c}_i$). The Fuzzy C-means algorithm is presented in Algorithm 3.2.

Fuzzy C-means can be considered as an extension of K-means where one observation can belong to multiple clusters. Therefore, the same validity criteria for identifying the number of clusters can be used. The two main advantages of FCM over other clustering algorithms is that outliers and noisy data can be handled [26, 38, 44, 50]. Basic FCM and advanced versions such as credibilistic FCM, Kernel FCM and distance-based FCM have been proposed for clustering multimode processes data [9, 16, 38, 40, 44, 50]. Similar to K-means, the main issue with applying Fuzzy C-means to mode identification is that transition mode clusters cannot be well separated from steady mode clusters.

Algorithm 3.2 Fuzzy C-means clustering algorithm

{Inputs: D: data set, k: number of clusters, ε: convergence threshold, $maxit$: maximum number
of iterations, ω: degree of fuzziness}
{ Initialization: $it = 0$, random initialization of the fuzzy partition matrix U }
repeat
   $it = it + 1$
   $C_i = \emptyset$ for $i = 1, 2, ..., k$
   {Centroid update step}
   for all $i = 1, 2, ..., k$ **do**
$$\mathbf{c}_i = \frac{\sum_{j=1}^{n} \mathbf{u}_{ij}^{\omega} \mathbf{x}_j}{\sum_{j=1}^{n} \mathbf{u}_{ij}^{\omega}}$$
   end for
   Calculate all distances according to Eq. (3.20)
   {Fuzzy partition matrix update step}
   for all $i = 1, 2, ..., k$ **do**
     for all $j = 1, 2, ..., n$ **do**
$$u_{ij}^{(it)} = \frac{1}{\sum_{l=1}^{k} (d_{ij}/d_{lj})^{2/(\omega-1)}}$$
     end for
   end for
until $\| U^{(it)} - U^{(it-1)} \|^2 \leq \varepsilon$ **or** $it = maxit$

3.4.3 Mixture Modeling Clustering

Given n_n observations in a m dimensional space ($D_n = \{\mathbf{x}\}_{j=1}^{n_n}$, $\mathbf{x}_j \in \Re^m$) and a
number of groups k to cluster the data, the objective of the Mixture modeling (MM)
clustering algorithm is to form the set of groups $C = \{C_1, C_2, ..., C_k\}$, such that the
following optimization criterion is maximized

$$P(D|\theta) = \prod_{j=1}^{n} f(\mathbf{x}_j) \tag{3.21}$$

where $P(D|\theta)$ represents the conditional probability of the data set D given the
parameters θ of a predefined statistical model determined by the function $f(\mathbf{x}_j)$.
This problem formulation is known as maximum likelihood estimation. Considering
a Gaussian mixture model (each cluster has a multivariate Gaussian distribution) the
function $f(\mathbf{x}_j)$ is characterized as follows

$$f(\mathbf{x}_j) = \sum_{i=1}^{k} f(\mathbf{x}_j|\mu_i, \Sigma_i) P(C_i) \tag{3.22}$$

$$f(\mathbf{x}_j|\mu_i, \Sigma_i) = \frac{1}{(2\pi)^m |\Sigma_i|^{1/2}} \exp -\frac{(\mathbf{x}_j-\mu_i)^T \Sigma_i^{-1}(\mathbf{x}_j-\mu_i)}{2} \tag{3.23}$$

and the set of all the model parameters are compactly written as

$$\theta = \{\mu_1, \Sigma_1, P(C_1), ..., \mu_k, \Sigma_k, P(C_k)\} \tag{3.24}$$

where μ_i is the mean, Σ_i is the covariance matrix, and $P(C_i)$ is the mixture probability of each cluster i such that $\sum_{i=1}^{k} P(C_i) = 1$. Therefore, the optimization criterion of Eq. 3.21 is reformulated as follows

$$ln P(D|\theta) = \sum_{j=1}^{n} ln(\sum_{i=1}^{k} f(\mathbf{x}_j|\mu_i, \Sigma_i)P(C_i)) \qquad (3.25)$$

where the *log-likelihood* operator is applied for simplification purposes [48].

The expectation-maximization (EM) algorithm [22] is used for finding the maximum likelihood estimates of the parameters θ. Given an initial estimate for θ, EM is a two-step iterative method: expectation and maximization. The cluster posterior probabilities $P(C_i|\mathbf{x}_j)$ are computed in the expectation step via the Bayes theorem, and the parameters θ are re-estimated in the maximization step. The EM algorithm for MM clustering is presented in Algorithm 3.3. The Matlab© implementation of this algorithm is shown in Appendix D.2.

Algorithm 3.3 Expectation-maximization algorithm for clustering

---

{Inputs: D: data set, k: number of clusters, ε: convergence threshold, *maxit*: maximum number of iterations}

{ Initialization: $it = 0$, random initialization of k centroids, $\Sigma_i^{(it)} = \mathbf{I}$ and $P(C_i)^{(it)} = 1/k \ \forall i = 1, ..., k$ }

repeat

    $it = it + 1$

    $C_i = \emptyset$ for $i = 1, 2, ..., k$

    {Expectation step}

    for all $i = 1, 2, ..., k$ and $j = 1, 2, ..., n$ **do**

        $w_{ij} = \dfrac{f(\mathbf{x}_j|\mu_i, \Sigma_i)P(C_i)}{\sum_{a=1}^{k} f(\mathbf{x}_j|\mu_a, \Sigma_a)P(C_a)}$

    end for

    {Maximization step}

    for all $i = 1, 2, ..., k$ **do**

        $\mu_i^{(it)} = \dfrac{\sum_{j=1}^{n} w_{ij}\mathbf{x}_j}{\sum_{j=1}^{n} w_{ij}}$ {Update mean}

        $\Sigma_i^{(it)} = \dfrac{\sum_{j=1}^{n} w_{ij}(\mathbf{x}_j - \mu_i)(\mathbf{x}_j - \mu_i)^T}{\sum_{j=1}^{n} w_{ij}}$ {Update covariance}

        $P(C_i)^{(it)} = \dfrac{\sum_{j=1}^{n} w_{ij}}{n}$ {Update priors}

    end for

until $\sum_{i=1}^{k} \| \mu_i^{(it)} - \mu_i^{(it-1)} \|^2 \le \varepsilon$ **or** $it == maxit$

The MM algorithm and some extensions have been widely used for mode identification in multimode processes [11, 17–19, 34, 35, 45–47, 51]. While the classic approach to clustering by using MM considers that each cluster has a multivariate normal distribution, some extensions have been proposed to work with clusters that may have non-Gaussian distributions [4], to deal with outliers [18] and missing data [28].

Similar to the K-means algorithm, the main parameter to be set is the number of clusters (modes) to be identified in the data set. Some alternative validity criteria commonly used are the Minimum Message Length criterion [39], the Bayesian Ying-Yang learning-based criterion [5, 43], the Bayesian information measure [34], Kernel Density Estimation on score variables [18, 46], and entropy penalized operators [19]. However, few MM methods extensions can identify transitions between steady modes. Two exceptions are the proposal presented in Ref. [18], and the formulation of the EM algorithm for identifying parity space-based residual generators taking into account that the process is a dynamic system [10, 29].

3.4.4 Window-Based Clustering

Window-based methods (WB) are based on hierarchical algorithms that discover a set of clusters by considering the order in which the observations have been acquired as an important factor. Since data from multimode processes are time series, window-based methods have the potential to identify the number of clusters (modes) automatically. Moreover, clusters that correspond to transition modes can be separated from clusters that correspond to steady modes.

The general structure of window-based algorithms is described as follows. Given n_n observations in a m dimensional space ($D_n = \{\mathbf{x}\}_{j=1}^{n_n}, \mathbf{x}_j \in \Re^m$), the observation space is crisply partitioned into L continuous sets where each one is called a window and it is formed by T_l observations ($W_l = \{\mathbf{x}\}_{j=1}^{T_l}$). The space partition is obtained based on the maximization of the following optimization criterion

$$J(T) = \sum_{i=1}^{k} \sum_{l=1}^{L} u_{il} \delta(\mathbf{s}_i, \mathbf{s}_l) \tag{3.26}$$

where $u_{il} \in [0, 1]$, $u_{il} = 1$ if the window W_l is a member of cluster C_i, otherwise $u_{il} = 0$ $\delta(\mathbf{s}_i, \mathbf{s}_l)$ is a similarity metric, $\mathbf{s}_i$ and $\mathbf{s}_l$ are two representative feature vectors of the cluster C_i and window W_l respectively.

The algorithm to perform window-based clustering is shown in Algorithm 3.4. The Matlab© implementation of this algorithm is shown in Appendix D.3. The different variants of window-based clustering algorithms depend on the similarity metric δ and the feature vector $\mathbf{s}$. For instance, the Euclidean distance with the maxima of the probability density function calculated for each window has been used [24].

The window length is the most important parameter to be set for these methods. Different partitions of the observation space can be obtained depending on the window size. The selection of this parameter depends on the dynamic of the process taking into account a compromise between computational complexity and accuracy in the isolation of operating modes. If the length of the window is too long the detection of mode changes is not precise. On the contrary, if a small window length is selected, the influence of noise could lead to the division of a stable mode including false transition states.

Window-based approaches allow to obtain the number of modes automatically and to identify transitions as well as steady modes [2, 13, 14, 24, 31–33, 52]. However, applying moving window methods can be difficult because selecting a similarity measure and designing a strategy for distinguishing transitional from steady modes are not easy tasks. Most works distinguish steady modes from transition modes based on the duration (number of sequential observations of each group) [13, 24, 31, 33] or the use of visualization tools [14, 53]. These criteria may be feasible to use from a practical standpoint as long as enough process knowledge is available.

Algorithm 3.4 Window-based clustering algorithm

{Inputs: D: data set, T: window length, γ: assignment threshold}

{ Initialization: $it = 0$, split the data set sequentially into L windows $\mathbf{W} = \{W_{l=1}^{L}$ of size T (the last set may have less than T observations), $C_j = W_l$ for $j = l = 1, 2, ..., L$, $k = L$, calculate the features $\mathbf{s}_i$ of each cluster}

Scale the variables

$NoClust^{(it)} = k$

{Step 1: separate transitions from steady modes}

repeat

  $it = it + 1$

  for all $i = 1, 2, ..., k$ **do**

    for all $l = 1, 2, ..., L$ **do**

      if $l == i + 1$ **and** $\delta(\mathbf{s}_i, \mathbf{s}_l) < \gamma$ **then**

        $C_i = C_i \bigcup \{C_l\}$

        {Update feature vector of cluster $\mathbf{s}_i$ }

        $\mathbf{C} = \mathbf{C} \setminus C_l$

        $L = L - 1, k = k - 1$

      end if

    end for

  end for

  $NoClust^{(it)} = k$

until $NoClust^{(it)} == NoClust^{(it-1)}$

Separate clusters of transitions $Ctrans$ from steady modes $Cstead$ according to some criteria

{Step 2: cluster steady modes without considering time information}

$it = 0$

$NoSteadClust^{(it)} = |Cstead|$

{Create feature vector for each steady cluster $\mathbf{sstead}_i^{(it)}$}

repeat

  $it = it + 1$

  for all $i = 1, 2, ..., k$ **do**

    $\delta_i = \min_l \{\delta(\mathbf{sstead}_i, \mathbf{sstead}_l)\}$ {Find similarity for the closest cluster}

    $minimum(i) = \arg\min_l \{\delta(\mathbf{sstead}_i, \mathbf{sstead}_l)\}$ {Find the index of the closest cluster}

  end for

  if $\min_i \delta_i < \gamma$ **then**

    $j = \arg\min_i \delta_i$

    $Cstead_j = Cstead_j \bigcup \{Cstead_{minimum(j)}\}$ {Merge steady modes}

    {Update feature vector of steady mode cluster $\mathbf{sstead}_j$ }

    $Cstead = Cstead \setminus Cstead_{minimum(j)}$ {Remove $Cstead_{minimum(j)}$}

  end if

  $NoSteadClust^{(it)} = |Cstead|$

until $NoSteadClust^{(it)} == NoSteadClust^{(it-1)}$

3.5   Clustering Data of Multimode Processes

In this section, it will be illustrated how to apply the proposed procedure to the nominal/normal data from the case studies presented in the previous chapter.

3.5.1   Continuous Stirred Tank Heater

The Continuous Stirred Tank Heater (CSTH) considers five operating modes, as described in the previous chapter, which are unknown. Even if the duration of the transition modes is short in comparison to the steady modes, and the number of variables is small, it is difficult to visually identify the total number of operating modes of the process. Therefore, the clustering procedure will be applied.

Since only three variables are measured, and in the same units (mA) there is no need to perform feature selection or apply preprocessing methods such as data scaling. Thus, the raw data will be used as inputs for the clustering methods. Three evaluation measures will be used to select the parameters of the clustering methods: Silhouette coefficient, Calinski-Harabasz index, and Davies-Bouldin index. Finally, the three clustering methods to be used are K-means, Mixture Modeling (MM), and window-based clustering.

3.5.1.1   K-Means Clustering

The parameter to be selected for the K-means clustering method is the number of clusters. Since the resulting clusters depend on the initialization of the algorithm multiple runs of the method are required. Therefore, the method is run a number of times. Here are considered five runs. The clustering evaluation measures determine how many clusters are identified in the data set. Thus, once a range of possible number of clusters is selected the trend in the evaluation measures must be observed in order to determine if a higher number of clusters provides a better performance. A range of clusters between two and ten is then selected. This range is defined based on expert knowledge, by considering the intuition that it is unlikely that there are more than ten operating modes for this process.

The results obtained are shown in Fig. 3.3. The best clustering result is found for the maximum numeric value of the Silhouette coefficient and Calinski-Harabasz, as well as the minimum numeric value of the Davies-Bouldin index. All indices present a similar trend. Above five clusters the evaluation measures show a lower performance. This indicates that there is no need to test a wider range of number of clusters. A Silhouette coefficient close to 1 is achieved with four and five clusters. The maximum of the Calinski-Harabasz index is achieved for four clusters while the minimum of the Davies-Bouldin index is achieved for five clusters in three out of the five runs. Therefore, it is not easy to infer the true number of clusters, being

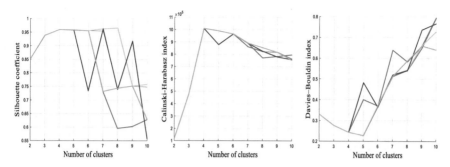

Fig. 3.3 Evaluation measures for different parameters (number of clusters)of the K-means clustering method with CSTH data

therefore considered four or five because the results are not conclusive. The reason is probably that the observations of the transitions affect the performance of the clustering method used.

3.5.1.2   MM Clustering

Similar to the K-means method, the parameter to be selected for the MM clustering method is the number of clusters. Moreover, the random initialization of k centroids also influences the resulting clusters. Therefore, the method is also run for five times considering the range of clusters between two and ten.

Figure 3.4 shows the three evaluation measures considering the MM clustering approach. In general, all indices present a similar trend, indicating that above four or five clusters the performance of the evaluation measures does not increase. Thus, there is no need to analyze a wider range for the number of clusters.

The Silhouette coefficient indicates the maximum by a slight difference for four clusters. Nonetheless, it is difficult to make conclusions from the results of the Silhouette coefficient because it is also close to one over the five runs with five to seven clusters. The Calinski-Harabasz index, on the other hand, achieves its maximum value for five clusters in three out of five runs. In the other two runs the maximum is achieved for four clusters. Finally, the Davies-Bouldin index has its minimum value for five clusters. By analyzing the results of the three indexes one might conclude that the number of clusters may be either four or five. It is difficult then to infer the true number of clusters. Again, the reason is probably that the observations of the transitions affect the performance of the clustering method used.

3.5.1.3   Windows-Based Clustering

Instead of setting the number of clusters to estimate, this method naturally discovers the number of clusters by adjusting the following parameters

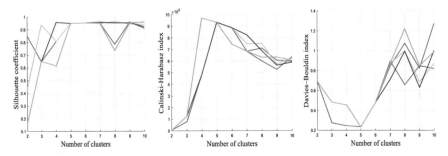

Fig. 3.4 Evaluation measures for different parameters (number of clusters) of the MM method with CSTH data

- Window length range: $T = [40, 220]$. The minimum window size is selected according to the time constants of the process. The window size is increased in 20 observations for each evaluation. The maximum analyzed can be modified if the evaluation measures indicate that a higher number of clusters present a better performance.
- The similarity metric (δ) used is the Euclidean distance. In the context of data from multimode processes, other similarity metrics do not provide a significant advantage.
- The feature vector (**s**) considered for each cluster is the mean vector. This selection is based on the fast computation time. If the results obtained at the end of the experiments offer no feasible conclusions regarding the appropriate number of clusters, then the maximum of the PDF can be used as the feature vector.
- Assignment threshold γ. This parameter represents the threshold to decide if the data of two consecutive windows are integrated or not. Therefore, it depends on both the similarity metric and the feature vector used for each window. A rule of thumb to select γ is to compute the median value of the similarity among all feature vectors and select around 5–10% of this value. For the CSTH process data considered, and the other parameters selected above yields, $\gamma = 0.05$.

Figure 3.5 shows the three evaluation measures which are computed only for the steady modes obtained through the window-based clustering method (WB). Thus, they evaluate the separability among the steady modes identified. Note that in this case multiple runs are not required. The number of steady modes identified for all window lengths is five. The best separability of the transition data from the steady mode clusters is achieved with a window length of 80 observations according to both the Silhouette coefficient and the Davies-Bouldin index. The maximum of the Calinski-Harabasz index is achieved for 40 observations. The main advantage of window-based clustering is the automatic discovery of the number of steady mode clusters, and the possibility of separating the transition from the steady modes. The estimation of the best window length helps to separate the steady modes from the first and last segments of the transitions. Varying the window length may affect also the number of steady modes discovered for processes with transitions of long duration.

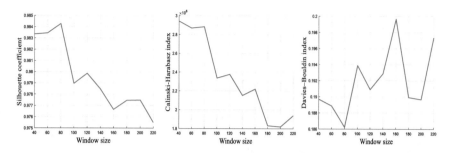

Fig. 3.5 Evaluation measures for different parameters of the WB method with CSTH data

However, the results obtained do not suggest that increasing the window size will allow to achieve a better performance in the evaluation measures because greater window sizes make them worst.

3.5.1.4   Partial Conclusions

Three clustering methods were applied to the data from the CSRH process in order to cluster the observations belonging to each operating mode. Although the K-means and MM methods suggest a number of steady modes which is close to the real one (four or five clusters) these methods do not allow to separate the steady data from the transitions. When the WB method was applied, the true number of operating modes (clusters) is always identified, with different degrees of separability. Therefore, for processes with fast transitions, it is suggested to use WB clustering methods.

3.5.2   Hanoi Water Distribution Network

The variations in the operating modes of the Hanoi WDN depend on the consumption patterns for this gravity-fed network. Although a different consumption pattern can be visually distinguished from weekdays to weekends, it is difficult to separate all consumption patterns from the entire data set. Moreover, the variability presents a non-steady behavior over the day. Therefore, the clustering procedure previously presented will be applied.

Initially the features to use as input for the clustering method are the raw variables. Since the measurement units of the variables are different, the min-max scaling is applied.

Since only three variables are measured in the same units (mA) there is no need to perform feature selection or apply preprocessing methods such as data scaling. Thus, the raw data will be used as inputs for the clustering methods. Three evaluation measures will be used to select the parameters of the clustering methods: Silhouette coefficient, Calinski-Harabasz index, and DaviesBouldin index. The clustering methods to be used are K-means, MM, and Window-based clustering.

3.5.2.1   K-Means Clustering with the Raw Data

The true number of modes for this process depends on the consumption patterns, which are three. However, how these consumption patterns affect the data is unknown in advance. Similar to the previous case study, the number of modes is unknown in advance. The range of number of clusters considered is then from two to ten clusters. This range is selected based on the idea that it is unlikely that more than ten different consumption patterns exist in such small network. If the evaluation measures indicate that the higher number of clusters present the best performance then this range should be extended, and a large number of clusters must be analyzed.

The results of the five runs of the algorithm are shown in Fig. 3.6. The Silhouette coefficient and the Davies-Bouldin index indicates that the appropriate number of clusters is two. The Calinkski-Harabasz index indicates ten clusters as appropriate. The trend of the three evaluation measures does not allow to arrive to conclusions with respect to the number of modes. The main reason is that the Calinski-Harabasz index indicates that more clusters allow to achieve better performance. This is contradictory with the other two evaluation measures. In fact, none of the evaluation measures indicates the true number of periodic patterns generated.

3.5.2.2   MM Clustering with the Raw Data

Similar to the previous algorithm, the range of number of clusters considered is also from two to ten. The results of the five runs of the algorithm are shown in Fig. 3.7. The Silhouette coefficient indicates as possible number of clusters two or three. The

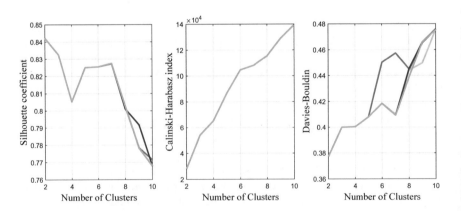

Fig. 3.6 Evaluation of the K-means clustering method with three indexes for HWDN case study

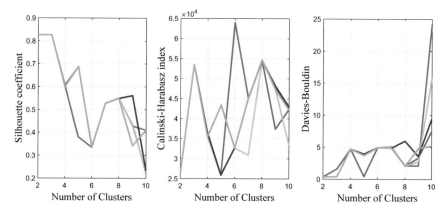

Fig. 3.7 Evaluation of the MM clustering method results with three indexes for HWDN case study

Calinkski-Harabasz index indicates three, six, and eight clusters in different runs of the algorithm. Finally, the Davies-Bouldin index indicates two, three, and five as possible number of clusters. Once again, it is not possible to draw any conclusion based on the results from the three evaluation measures in different runs of the algorithm.

3.5.2.3 WB Clustering with the Raw Data

The application of the window-based clustering method did not allow to arrive at any conclusion either. Independently of the different set of parameters such as window size and assigment threshold γ. The number of clusters identified was always identical to the number of windows initially defined when the value of γ was set very low. Conversely, when γ is increased the resulting number of clusters obtained is one.

3.5.2.4 Feature Selection for Improving Clustering Performance

The reason behind the poor results obtained is that the multimode behavior is not captured by the raw variables/time-series signals. In this case study, the difference among the different operating modes can be more clearly analyzed by considering the day to day variability caused by the periodic component of the demand.

The periodic component causes a non-stationary behavior of the variables over the day which difficults the separation of the different consumption patterns. Therefore, a more appropriate strategy is to search for the number of modes according to the day to day variability. Mathematically, the re-ordering of a data set formed by n observations and m variables can be represented as follows

$$D \in \mathfrak{R}^{n \times m} \rightarrow \{\hat{D}_1 \hat{D}_2 ... \hat{D}_{n_h}\} \mid \hat{D}_i \in \mathfrak{R}^{n_d \times m} \; and \; n = n_h \times n_d \qquad (3.27)$$

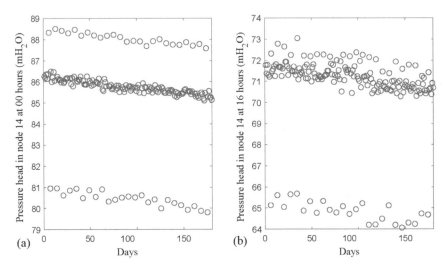

Fig. 3.8 Day to day variability of the pressure head behavior for node 14 at **a** 00:00 h and **b** 16:00 h

where n_h is the number of observations considered in one day, and n_d is the number of days that form the entire data set. Therefore, new data sets of features are created to perform the clustering task. This implies that the fault diagnosis task must be solved for each data set.

For the HWDN case study, there are six months of data, and assuming that the periodic behavior of the consumers has a duration of one day, a total of $n_d = 180$ days are approximately derived. Moreover, since the sampling time is 30 min, then $n_h = 48$ is the total number of observations for each day. The day to day observation variability for the observation at 00:00 and 16:00 h is shown in Fig. 3.8. The day to day variability allows to visually distinguish three and two clusters, respectively. Each pattern corresponds to the weekday and weekends. The effect of seasonal trend can also be visualized. The three periodic components overlap during some hours thus the number of clusters will vary between two an three depending on the hour of the day.

3.5.2.5   K-Means and MM Clustering with the Data Sets of Features

The K-means and MM clustering methods with the validity indexes are applied to each feature data set $\hat{D}_i$ where $i = \{1, ..., 48\}$. The optimal number of clusters according to each evaluation measure for each sampling time is shown in Fig. 3.9 for both the K-means and the MM clustering methods. It can be observed that depending on the time of the observation analyzed the number of clusters varies mostly between two and three. This occurs because each observation corresponds to a different time of the day, and the consumption patterns vary from day to day depending on the hour.

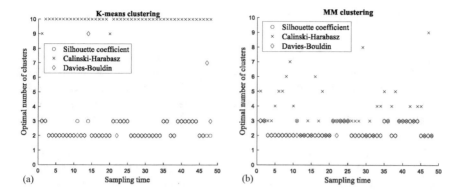

Fig. 3.9 Best number of clusters for each sampling time considering three evaluation measures with **a** the K-means and **b** the MM clustering methods

If a voting scheme is considered, based on the three evaluation metrics, the best number of clusters always remains between two and three independently of the clustering method. Here, it can be assumed that the number of clusters discovered represent steady modes because of the properties of the signals. The consequences of using this data re-ordering approach for fault diagnosis will be further analyzed in the next chapter.

Finally, it must be remarked that the window-based clustering method should not be applied in this case because the reorganization of the data set implies that the observations are not continuously sampled in time anymore. Therefore, the features extracted for each window would not be representative of the time correlations among the variables.

3.5.2.6   Partial Conclusions

Three clustering methods were applied to the raw data from the HWDN case study to cluster the observations belonging to each operating mode. Performing the clustering methods on the raw data set did not allow to arrive to any conclusions regarding the number of modes. When a feature transformation strategy is applied, then the true number of modes can be recovered. The feature transformation is tailored to the specific process that is considered meaning that some expert knowledge is required. This illustrates that in some cases the analysis of the raw data does not allow to extract any useful information for later performing fault diagnosis.

3.5.3   Tennessee Eastman Process

As shown in Table 2.4, the TEP is designed to operate in six different modes. Here
the modes 1, 2, and 3 are considered. In this case study the transition modes are long
and may last for hours given the slow dynamics of the process. However, given the
high number of variables, 31, it is difficult to visually identify or separate the data
corresponding to the different operating modes. Therefore, the clustering methods
studied in this chapter will be applied.

The scaling method selected for preprocessing the data is min-max. This is
required because the variables have different units. Initially, the raw data will be
used as inputs for the clustering methods. If no conclusions can be obtained from
the application of the clustering methods, then feature selection methods should
be considered. Three evaluation measures will be used to select the parameters of
the clustering methods: Silhouette coefficient, Calinski-Harabasz index, and Davies-
Bouldin index. Finally, the three clustering methods to be used are K-means, MM,
and WB.

3.5.3.1   K-Means Clustering

The parameter to be selected for the K-means clustering method is the number of
clusters. Since the resulting clusters depend on the initialization of the algorithm
multiple runs of the method are required. Therefore, the method is run for a number
of times. Here five runs are considered. The clustering evaluation measures determine
how many clusters are identified in the data set. Thus, once a range of possible number
of clusters is selected, the trend in the evaluation measures must be observed in order
to determine if a higher number of clusters provides a better performance. A range
of clusters between two and ten is then selected again based on the assumed possible
number of operating modes. If the evaluation measures indicate the best performance
for ten clusters, then a larger number of clusters should be analyzed.

The results are shown in Fig. 3.10. The Silhouette coefficient indicates that the
appropriate number of operating modes is five. The Calinski-Harabasz index indicates ten operating modes and possibly more if a wider range would be analyzed.
The Davies-Bouldin index indicates that the number of operating modes is four.
Therefore, it is not possible to arrive to any conclusion with respect to the number
of operating modes, neither there is the possibility of distinguishing steady modes
from transitions.

3.5.3.2   MM Clustering

The parameter to be selected for the MM clustering method is also the number of
clusters. The method is also run for five times considering the range of clusters
between two and ten based on the same criteria applied for the K-means clustering

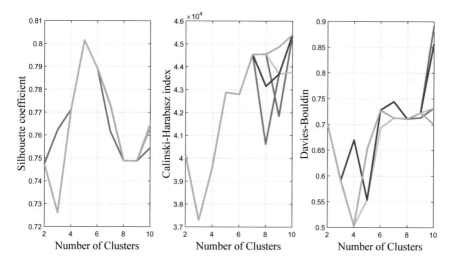

Fig. 3.10 Evaluation measures for different parameters (number of clusters) of the K-means clustering method with TEP data

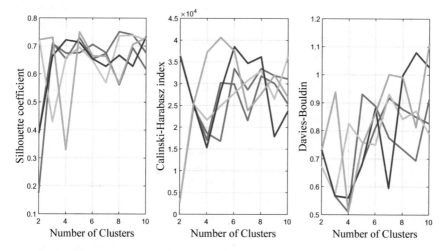

Fig. 3.11 Evaluation measures for different parameters (number of clusters) of the of MM clustering method with TEP data

algorithm. Figure 3.11 shows the three evaluation measures considering the MM clustering approach. The Silhouette coefficient as well as the Calinski-Harabasz index do not provide any useful information about the number of clusters, except for the fact that two clusters present the worst performance. The Davies-Bouldin index indicates that four clusters provides the best performance in four out of the five runs. However, it is not possible to arrive to any conclusion about the number of modes by analyzing the three indices.

3.5.3.3   Window-Based Clustering

The WB clustering method will be applied considering the following parameters:

- Window length range: $T = [40, 440]$. The minimum window size is selected according to the time constants of the process. The window size is increased in 40 observations for each evaluation. The maximum analyzed can be modified if the evaluation measures indicate that a higher number of clusters present a better performance.
- The similarity metric (δ) used is the Euclidean distance.
- The feature vector (**s**) considered for each cluster is the mean vector. If the results obtained at the end of the experiments offer no feasible conclusions regarding the appropriate number of clusters, then the maximum of the PDF can be used as feature vector.
- Assignment threshold γ. This parameter represents the threshold to decide if the data of two consecutive windows are integrated or not. Instead of using the rule of thumb presented in the CSTH case study, a grid search will be used by considering the range $= [0.05, 0.5]$ (intervals of evaluation equal to 0.05).

Overall, a grid search will be performed by considering the assignment threshold together with the window size because these two parameters determine the number of steady clusters estimated. For each assignment threshold the optimal window size depends on the evaluation measure. For illustration purposes, the evaluation measures corresponding to different window sizes and assignment thresholds of 0.05, 0.1 and 0.5 are shown in Table 3.1. It is highlighted in boldface the maximum or minimum of each evaluation measure that indicates the best partitioning of the data set in each experiment. The number of steady modes (No s.m.) is also shown for each combination of parameters.

The Silhouette measure does not allow to distinguish between 3 and 4 modes for assignment thresholds of 0.05 and 0.1. When the assignment threshold increases to 0.5 the Silhouette measure gets worst from 0.99 to 0.94. The best values of the Davies-Bouldin index and the Calinski-Harabasz index are 0.1310 and 618090, respectively. The Davies-Bouldin index suggests that four steady modes achieve the best separability with a window size of 240 observations. The Calinski-Harabasz index suggests three steady modes with a window size of 400 observations. If a voting scheme is applied across the three experiments then the number of modes selected is three. However, if the maximum performance is considered for each evaluation measure then four modes are selected. Therefore, there are no concluding results regarding the number of clusters except that the true number lies between three and four.

The partitioning of the data into steady modes and transitions is shown in Fig. 3.12 for each set of parameters. Every mode change is well identified with the window size of 240 observations. The first steady mode simulated lasts for 1000 observations so it cannot be distinguished with the window size of 400 because the minimum length in this case is 1200 observations. When a window size of 240 is selected, the first mode simulated is identified as a different mode from the last one even when the process operates under the same steady mode. This happens because the features

Table 3.1 Evaluation measures for different window sizes T and assignment thresholds γ for the TEP study case

Assignment threshold of 0.05

Window size	40	80	120	160	200	240	280	320	360	400
No s.m.[a]	13	8	7	5	4	4	3	3	3	3
Sil	0.23	0.42	0.44	0.70	**0.99**	**0.99**	**0.99**	**0.99**	**0.99**	**0.99**
CH	133777	214113	218533	358670	445226	417664	522117	514988	475899	**523589**
DB	2.52	2.41	2.61	1.10	0.1315	**0.1310**	0.1340	0.1344	0.1345	0.1341

Assignment threshold of 0.1

No s.m.[a]	31	6	5	5	4	4	3	3	3	3
Sil	0.45	0.76	0.82	0.80	**0.99**	**0.99**	**0.99**	**0.99**	**0.99**	**0.99**
CH	67912	317326	381577	371889	502922	501350	607688	597917	536982	**618090**
DB	1.46	0.77	0.80	0.98	**0.1352**	0.1357	0.1385	0.14078	0.1413	0.1383

Assignment threshold of 0.5

No s.m.[a]	17	14	11	8	7	7	4	3	3	3
Sil	0.74	0.79	0.80	0.81	0.79	0.80	0.89	**0.94**	**0.94**	**0.94**
CH	42496	55705	68634	95854	95118	102081	181678	223213	221938	**235896**
DB	0.76	0.71	0.60	0.67	0.67	0.63	0.48	0.2676	0.2614	**0.2609**

[a] Number of steady modes

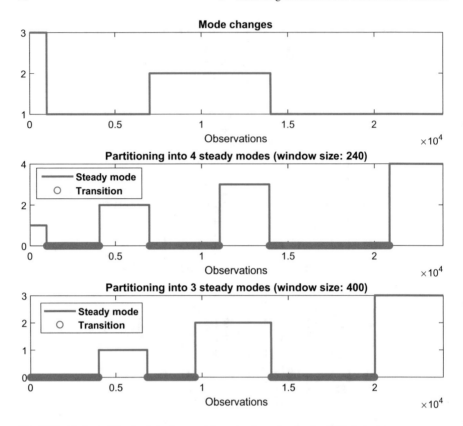

Fig. 3.12 Mode partitioning into three and four steady modes for the TEP study case

of the variables are different. However, this issue does not impact the fault diagnosis performance significantly. In the next chapter, the two possible clustering alternatives (three and four steady modes) will be used to build the fault detection approach.

3.5.3.4   Partial Conclusions

Three clustering methods were applied to the data from the TEP to cluster the observations belonging to each operating mode. The K-means and MM methods do not allow to arrive to any conclusions regarding the number of modes. Conversely, when the window-based method was applied the number of operating modes identified (clusters) is either three or four, when the true number is three. The main reason for this result is that both K-means and MM methods generally work well for convex shaped clusters. Thus, as the data set includes transitions with long duration the shape of the clusters is non-convex. In these cases, it can be recommended then to use the window-based clustering approach (WB).

Remarks

Clustering is an important task to be performed, in order to identify the features of multimode processes: number of modes, mode type, and observations corresponding to each mode. The features of the data are different depending on the multimode process. Therefore, identifying the multimode features is not a simple task, in which some degree of expert knowledge plays a fundamental role. If the multimode process involves transitions, then traditional clustering methods such as K-means and MM clustering are not useful to separate steady modes from transitions. Therefore, Window-based clustering is preferred in this case. In other processes such as water distribution networks, the multimode characteristic is not discovered by analyzing the raw data. Some preprocessing is then required for cases like this one. The application of three of the most used clustering algorithms and evaluation measures was made to the three cases of study described in the previous chapter.

References

1. Aggarwal, C.C., Reddy, C.K.: Data Clustering. Chapman & Hall/CRC, New York, USA (2014)
2. Beaver, S., Palazoglu, A., Romagnoli, A.: Cluster analysis for autocorrelated and cyclic chemical process data. Ind. Eng. Chem. Res. **46**(11), 3610–3622 (2007)
3. Calinski, T., Harabasz, J.: A dendrite method for cluster analysis. Commun. Stat. **3**(1), 1–27 (1974)
4. Chen, J., Yu, J.: Independent component analysis mixture model based dissimilarity method for performance monitoring of non-Gaussian dynamic processes with shifting operating conditions. Ind. Eng. Chem. Res. **53**(13), 5055–5066 (2014)
5. Choi, S.W., Martin, E.B., Morris, A.J., Lee, I.B.: Fault detection based on a maximum-likelihood principal component analysis (PCA) mixture. Ind. Eng. Chem. Res. **44**(7), 2316–2327 (2005)
6. Dash, M., Liu, H.: Feature selection for clustering. In: Terano, T., Liu, H., Chen, A.L.P. (eds.) Knowledge Discovery and Data Mining. Current Issues and New Applications, pp. 110–121. Springer, Berlin (2000)
7. Davies, D.L., Bouldin, D.W.: A cluster separation measure. IEEE Trans. Patt. Anal. Mach. Intell. **1**(2), 224–227 (1979)
8. Du, W., Fan, Y., Zhang, Y.: Multimode process monitoring based on data-driven method. J. Frankl. Inst. **354**, 2613–2627 (2017)
9. Ge, Z., Song, Z.: Multimode process monitoring based on Bayesian method. J. Chemom. **23**(12), 636–650 (2009)
10. Haghani, A., Jeinsch, T., Ding, S.X.: Quality-related fault detection in industrial multimode dynamic processes. IEEE Trans. Ind. Electr. **61**(11), 6446–6453 (2014)
11. Haghani, A., Krueger, M., Jeinsch, T., Ding, S.X., Engel, P.: Data-driven multimode fault detection for wind energy conversion systems. IFAC-PapersOnLine **48**(21), 633–638 (2015)
12. Han, J., Kamber, M., Pei, J.: Data Mining Concepts and Techniques. Elsevier (2012)
13. He, Y., Ge, Z., Song, Z.: Adaptive monitoring for transition process using dynamic mutual information similarity analysis. In: Chinese Control and Decision Conference, pp. 5832–5837 (2016)
14. He, Y., Zhou, L., Ge, Z., Song, Z.: Dynamic mutual information similarity based transient process identification and fault detection. Can. J. Chem. Eng. (2017). https://doi.org/10.1002/cjce.23102

15. Jain, A.K.: Data clustering: 50 years beyond K-means. Pattern Recognit. Lett. **31**, 651–666 (2010)
16. Liu, J.: Fault detection and classification for a process with multiple production grades. Ind. Eng. Chem. Res. **47**(21), 8250–8262 (2008)
17. Liu, J.: Data-driven fault detection and isolation for multimode processes. As. Pac. J. Chem. Eng. **6**(3), 470–483 (2011)
18. Liu, J., Chen, D.S.: Nonstationary fault detection and diagnosis for multimode processes. AIChE J. **56**(1), 207–219 (2010)
19. Ma, L., Dong, J., Peng, K.: Root cause diagnosis of quality-related faults in industrial multimode processes using robust Gaussian mixture model and transfer entropy. Neurocomputing **285**, 60–73 (2018)
20. Milligan, G.W., Cooper, M.C.: An examination of procedures for determining the number of clusters in a data set. Psychometrika **50**(2), 159–179 (1985). https://doi.org/10.1007/BF02294245. http://link.springer.com/10.1007/BF02294245
21. Mirkin, B.G.B.G.: Clustering for Data Mining: a Data Recovery Approach. Chapman & Hall/CRC (2005)
22. Moon, T.: The expectation-maximization algorithm. IEEE Signal Process. Mag. **13**(6), 47–60 (1996). https://doi.org/10.1109/79.543975
23. Natarajan, S., Srinivasan, R.: Multi-model based process condition monitoring of offshore oil and gas production process. Chem. Eng. Res. Des. **88**(5–6), 572–591 (2010)
24. Quiñones-Grueiro, M., Prieto-Moreno, A., Llanes-Santiago, O.: Modeling and monitoring for transitions based on local kernel density estimation and process pattern construction. Ind. Eng. Chem. Res. **55**(3), 692–702 (2016)
25. Quiñones-Grueiro, M., Prieto-Moreno, A., Verde Rodarte, C., Llanes-Santiago, O.: Data-driven monitoring of multimode continuous processes: a review. Chemom. Intell. Lab. Syst. **189**, 56–71 (2019)
26. Rodríguez-Ramos, A., Silva Neto, A.J., Llanes-Santiago, O.: An approach to fault diagnosis with online detection of novel faults using fuzzy clustering tools. Expert Syst. Appl. **113**, 200–212 (2018)
27. Rousseeuw, P.: Silhouettes: a graphical aid to the interpretation and validation of cluster analysis. J. Comput. Appl. Math. **20**, 53–65 (1987)
28. Sammaknejad, N., Huang, B.: Operating condition diagnosis based on HMM with adaptive transition probabilities in presence of missing observations. AIChE J. **61**(2), 477–493 (2015)
29. Sari, A.H.A.: Data-Driven Design of Fault Diagnosis Systems-Nonlinear Multimode Processes. Springer (2014)
30. Sivogolovko, E., Novikov, B.: Validating cluster structures in data mining tasks. In: Proceedings of the 2012 Joint EDBT/ICDT Workshops on-EDBT-ICDT '12, p. 245. ACM Press, New York, USA (2012). https://doi.org/10.1145/2320765.2320833. http://dl.acm.org/citation.cfm?doid=2320765.2320833
31. Song, B., Tan, S., Shi, H.: Key principal components with recursive local outlier factor for multimode chemical process monitoring. J. Process Control **47**, 136–149 (2016)
32. Srinivasan, R., Wang, C., Ho, W.K., Lim, K.W.: Dynamic principal component analysis based methodology for clustering process states in Agile chemical plants. Ind. Eng. Chem. Res. **43**(9), 2123–2139 (2004)
33. Tan, S., Wang, F., Peng, J., Chang, Y., Wang, S.: Multimode process monitoring based on mode identification. Ind. Eng. Chem. Res. **51**(1), 374–388 (2012)
34. Thissen, U., Swierenga, H., de Weijer, A., Melssen, W.J., Buydens, L.M.C.: Multivariate statistical process control using mixture modelling. J. Chemom. **19**(1), 23–31 (2005)
35. Tong, C., El-Farra, N.H., Palazoglu, A., Yan, X.: Fault detection and isolation in hybrid process systems using a combined data-driven and observer-design methodology. AIChE J. **60**(8), 2805–2814 (2014)
36. Tong, C., Yan, X.: Double monitoring of common and specific features for multimode process. As. Pac. J. Chem. Eng. **8**(5), 730–741 (2013)

37. Wang, F., Tan, S., Peng, J., Chang, Y.: Process monitoring based on mode identification for multi-mode process with transitions. Chemom. Intell. Lab. Syst. **110**(1), 144–155 (2012)
38. Wang, X., Wang, X., Wang, Z., Qian, F.: A novel method for detecting processes with multi-state modes. Control Eng. Pract. **21**, 1788–1794 (2013)
39. Xie, X., Shi, H.: Dynamic multimode process modeling and monitoring using adaptive Gaussian mixture models. Ind. Eng. Chem. Res. **51**(15), 5497–5505 (2012)
40. Xie, X., Shi, H.: Multimode process monitoring based on fuzzy c-means in locality preserving projection subspace. Chin. J. Chem. Eng. **20**(6), 1174–1179 (2012)
41. Xiong, H., Gaurav Pandey, Steinbach, M., Kumar, V.: Enhancing data analysis with noise removal. IEEE Trans. Knowl. Data Eng. **18**(3), 304–319 (2006)
42. Xu, R., Wunsch, D.C.: Clustering. IEEE Press (2009)
43. Xu, X., Xie, L., Wang, S.: Multimode process monitoring with PCA mixture model. Comput. Electr. Eng. **40**(7), 2101–2112 (2014)
44. Yoo, C.K., Vanrolleghem, P.A., Lee, I.B.: Nonlinear modeling and adaptive monitoring with fuzzy and multivariate statistical methods in biological wastewater treatment plants. J. Biotechnol. **105**(1-2), 135–163 (2003)
45. Yu, H.: A novel semiparametric hidden Markov model for process failure mode identification. IEEE Trans. Autom. Sci. Eng. **15**(2), 506–518 (2017)
46. Yu, J.: A nonlinear kernel Gaussian mixture model based inferential monitoring approach for fault detection and diagnosis of chemical processes. Chem. Eng. Sci. **68**(1), 506–519 (2012)
47. Yu, J.: A new fault diagnosis method of multimode processes using Bayesian inference based Gaussian mixture contribution decomposition. Eng. Appl. Artif. Intell. **26**(1), 456–466 (2013)
48. Zaki, M.J., Meira, W.: Data Mining and Analysis. Cambridge University Press (2014)
49. Zhang, S., Wang, F., Tan, S., Wang, S., Chang, Y.: Novel monitoring strategy combining the advantages of the multiple modeling strategy and Gaussian Mixture Model for Multimode Processes. Ind. Eng. Chem. Res. **54**(47), 11866–11880 (2015)
50. Zhang, S., Zhao, C.: Sationarity test and Bayesian monitoring strategy for fault detection in nonlinear multimode processes. Chemom. Intell. Lab. Syst. **168**, 45–61 (2017)
51. Zhu, J., Ge, Z., Song, Z.: Recursive mixture factor analyzer for monitoring multimode time-variant industrial processes Recursive mixture factor analyzer for monitoring multimode time-variant industrial processes. Ind. Eng. Chem. Res. **55**(16), 4549–4561 (2016)
52. Zhu, Z., Song, Z., Palazoglu, A.: Transition process modeling and monitoring based on dynamic ensemble clustering and multiclass support vector data description. Ind. Eng. Chem. Res. **50**(24), 13969–13983 (2011)
53. Zhu, Z., Song, Z., Palazoglu, A.: Process pattern construction and multi-mode monitoring. J. Process Control **22**, 247–262 (2012)

Chapter 4
Monitoring of Multimode Continuous Processes

Abstract This chapter presents three different monitoring schemes developed for multimode processes. An important challenge to be tackled consists on differentiating the change of mode from a fault. The complexity of this task is related to the types of modes of a process: steady and non-steady. Three different types of benchmark processes are considered.

4.1 Process Monitoring Loop for Multimode Processes

The classic data-driven monitoring loop presented in Chap. 1 considers as the first step the fault detection task. However, the use of conventional fault detection methods for multimode processes is not straightforward. The nominal data distribution of a multimode process is divided into different clusters such that in order to solve the fault detection task, the operating mode identification task must be solved first online. The operating mode identification task can be defined as follows

Definition 4.1 (*Online data-driven operating mode identification*) Given a set of observations $D_o = \{\mathbf{x}\}_{j=1}^{n_o}$, $\mathbf{x}_j \in \Re^m$, $D_f \subset D$), and a set of known operating modes $\Phi = \{op_1, op_2, ..., op_O\}$ design a function $f_c(D_o, \Phi)$: $D_o \in \Re^{n \times m} \to op_o \in \Phi$ that associates the data set D_o with one of the O possible operating modes.

As it can be observed, the operating mode identification task is framed as a pattern recognition problem. The introduction of this new task modifies the process monitoring loop according to Fig. 4.1. Therefore, the initial task allows to determine the operating mode before running the rest of the loop. This book concentrates on the strategies proposed to solve this first step coupled with the fault detection task. In the next section, the methodology to design methods for mode identification and fault detection is described.

© Springer Nature Switzerland AG 2021

M. Quiñones-Grueiro et al., *Monitoring Multimode Continuous Processes*,
Studies in Systems, Decision and Control 309,
https://doi.org/10.1007/978-3-030-54738-7_4

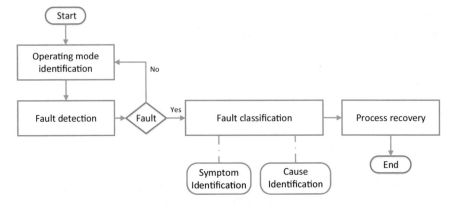

Fig. 4.1 Fault diagnosis/monitoring loop for multimode processes

4.1.1 Design of Operating Mode Identification and Fault Detection Methods

The solution of both the operating mode identification and fault detection problems is usually tackled together. Therefore, the following procedure can be applied based on the CRISP-DM methodology described in Chap. 1. (See Fig. 1.2).

1. The first step, called business understanding, is related to the characterization of the process and data collection. If possible, a data set which is representative of all operating conditions of a process must be acquired. In an industry with hundreds of variables, the variables to be collected must also be selected. Moreover, some degree of expert knowledge about the process to be monitored is very useful to characterize in advance some features of the process, such as dynamics, non-linearities, and others.
2. The second step, called data understanding, is related to the data-driven verification of the features of the process, and the quality of the collected data set. The algorithm presented in Chap. 2 to identify if a process is multimode or not is used in this step to verify this feature. Additional properties of the process according to the data set could be identified in this step such as non-linearity, and dynamic behavior.
3. The third step involves the data preparation task. The clustering procedure presented in Chap. 3 is applied here to separate the data set into different sets where each one corresponds to an operating mode of the process. This step and the next one are very much related, such that each strategy and method to be used for modeling may require a particular data processing.
4. The fourth step is modeling, and it is related to the design of the operating mode identification strategy as well as the fault detection method to be used. The solution of this step is the main focus of this chapter, and it will be presented in detail in the following sections.

5. The fifth step is evaluation, and it is related to evaluate the performance of the fault detection task (which depends on the correct mode identification).
6. The six step is related to the integration of the methods designed into a larger software system. Software and hardware specifications are important to properly integrate the methods designed into the industry.

4.1.2 Calibration of Operating Mode Identification and Fault Detection Methods

The steps four and five of the methodology are usually accomplished together, in order to calibrate the parameters of the methods used. The process of calibrating the parameters of the methods is known as *training*. Once the parameters of the method are selected, the process of testing its performance on a different data set is known as *testing*.

The execution time of the *training* and *testing* processes are different. The former one occurs off-line. This means that there are no time requirements for the execution of the method. Conversely, the latter occurs on-line. This means that the execution time of the fault diagnosis loop should be smaller than the sampling time of the process. For instance, in a process with a sampling time of five minutes, such as the control of temperature in a room, the execution time of the fault diagnosis loop should last less than five minutes. This depends, of course, on the hardware available for the implementation. This should be analyzed in step number six of the CRISP-DM methodology. It is not the purpose of this book to analyze the computational requirements for the implementation of the fault diagnosis methods. However, most of the tools presented have a small execution time for its on-line deployment.

The procedure for parameter estimation is based on a ten-fold cross validation [8]. The classifier parameters are readjusted in an iterative fashion such that the best possible accuracy is obtained for the training set. Finally, the classifiers' overall accuracy is estimated by using the test data. The Friedman and Wilcoxon non-parametric hypothesis tests are used to determine if there are statistically significant differences in performance among the different sets of parameters based on the cross-validation results [7].

4.2 Operating Mode Identification and Fault Detection Strategies

The number of modes, their types (steady or transition) and the observations belonging to each one of the modes are found first through clustering methods. Then, a mode identification and fault detection strategy must be designed to cope with the multimode behavior. The three main strategies to be developed are shown in Fig. 4.2.

The first group of strategies (Fig. 4.2a.1 and a.2) consider a single fault detection method. To deal with the multimode behavior two strategies can be used: (a.1) to transform the data to remove the multimode feature, and (a.2) to adapt the fault detection model as the mode changes. The second group of strategies consider a fault detection model for each operating mode (Fig. 4.2b.1 and b.2). To deal with the multimode behavior two strategies can be used: (b.1) design a function to map the observation to a specific operating mode, and (b.2) consider the probability of a fault by analyzing the results of all the fault detection models. Below each one of these strategies is described in detail.

4.2.1 Data Transformation and Adaptive Schemes for Fault Detection

The first mode identification strategy consists of coupling a data transformation or adaptive technique with a fault detection method. The transformation technique follows a nearest neighbor approach (Fig. 4.2a.1). Each new observation is scaled according to a set of data points by considering the nearest neighbors. Such transformation allows to ignore the mode clustering task, and it is effective as long as only steady modes are considered [37, 40, 43, 55, 56, 65, 83]. This approach has been further improved for dynamic processes by calculating the neighborhood, taking into account both the time and spatial information [56]. However, the computational cost involved in the calculation of the neighborhoods may be high depending on the size of the data set and the number of variables. In addition, selecting the number of neighbors and the similarity metric can become a very complex task, depending on some features of the data set such as outliers and noise.

The adaptive monitoring technique (Fig. 4.2a.2) relies in updating the parameters of the fault detection model according to the evolution of the process [11, 17, 35, 87]. One important decision to make before updating the parameters of the fault detection model is whether a fault is affecting the system or not. Since the fault detection is built based on nominal data, the updating with faulty data must be avoided. It is difficult to avoid the blind update of the parameters when the faults evolve very slowly. In fact, the fault smearing problem is widely known in the development of recursive and moving window strategies for adapting the parameters of classic fault detection models [34]. Some strategies to decide when to update are knowledge-based rules [30, 35], the manipulation of a forgetting factor [14, 24, 67], and a Bayesian approach [42, 70]. The use of the adaptive approach generally assumes only steady modes because classic fault detection methods are designed for steady conditions.

Fig. 4.2 Strategies for mode identification and fault detection

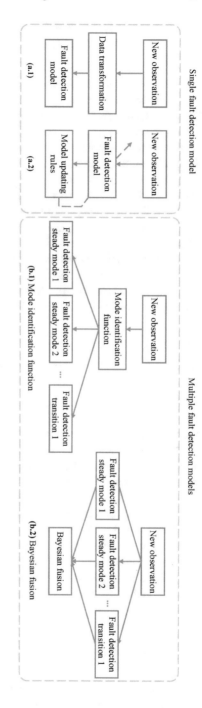

4.2.2 Mode Selection Function for Fault Detection with Multiple Models

The aim of a mode selection function (Fig. 4.2b.1) is to track the operating mode of the process to select the corresponding fault detection model [1, 10, 19, 26, 29, 39, 46, 47, 49, 66, 80]. Decoupling the mode identification from the fault detection step allows to use a different data-driven model for each operating regime, including steady modes and transitions [26, 49, 63]. Moreover, knowing the operating mode helps the process operator for decision making.

The mode selection function plays a primary role for fault detection because if the wrong mode is identified, the monitoring performance can significantly deteriorate. Two possible decision functions are based on: the model fitness to current data [19, 46, 49, 64, 80, 82, 85], and the references or set-points of controlled variables in industrial processes [15]. Some fitness measures are: the squared prediction error [82, 85], the number of nearest neighbors [19], probability functions [46, 64] and the Euclidean [45] or Mahalanobis distance [38] to the center of the cluster representing each mode. Hidden Markov Models can also be used for mode identification by employing the Viterbi Algorithm [1, 39, 52, 54, 62, 63, 66].

4.2.3 Bayesian Fusion of Multiple Models for Fault Detection

The Bayesian fusion approach combines the fault detection results of the model of each operating mode in a probabilistic way (Fig. 4.2b.2). The Bayesian fusion avoids the need of the above-mentioned mode identification function by assuming a soft mode identification: each observation belongs to a cluster with a certain degree. The fusion strategy should be independent of the fault detection model developed for each mode. Otherwise, there is no flexibility for adjusting the model of each mode [3, 4, 20, 21, 31, 41, 63, 73, 76, 84, 86].

A natural approach to develop this scheme is to use Mixture Models to characterize the data set [4, 31, 41, 76]. Other approaches formulate Bayesian inference systems merging the fault detection results of all the independent model [53, 63, 84, 86]. To do this, the posterior probabilities of each operation mode are computed. For example, the probability of the assignment to a specific mode corresponds to the weighted ratio of the fault detection indices together with their respective fault detection thresholds.

4.3 Fault Detection Methods

Traditionally, fault detection in the process industry has been synonymous of monitoring quality variables or some key process indicators based on Statistical Process Control methods [23]. Early fault detection methods monitor each variable independently (univariate analysis). As a result, it is possible to appreciate a large number of control charts in the control centers of industrial plants. If a fault occurs in the operation of the plant, several of these graphics announce alarms in a short period of time or simultaneously. This occurs because the process variables are correlated, and an abnormal event can affect many of these variables at the same time. When this situation occurs, it is very difficult for operators to isolate and determine the source of the problem, which can only be in one of the many variables in alarm status.

Multivariate statistical control techniques have provided a solution to the above-mentioned problem, since they also characterize the correlations among variables. Therefore, they are able to extract information useful for process monitoring more effectively. With these tools, a plant can be supervised by operators attending only a few control charts. These graphics are based on multivariate projection methods and they are easy to understand for operators, which is the reason for which they are widely used nowadays in the control center of industrial plants [32, 51].

Multimode processes may present two types of modes: steady and transition. The features of the data set corresponding to each type of mode are different. Most fault detection methods have been proposed for steady mode monitoring since industrial plants mostly operate under steady state conditions. However, most plants have transitions such as startups, shutdowns, and changes in the production mode. It is more challenging to monitor the processes during transitions. Therefore, specific methods tailored to transition monitoring have been proposed.

Figure 4.3 provides basic guidance to select the fault detection method according to the features of the data. As shown, the first division is based on the type of mode: steady or transition. Then, the fault detection model of steady modes depends on the features of the data set whereas the fault detection model of transitions depends primarily on the availability of data.

Overall, a wide variety of multivariate statistical techniques and artificial intelligence methods can be used for fault detection during steady modes. Each method is tailored to the features of the data. If the relations among the process variables are non-linear, then techniques such as Support Vector Data Description, Neural Networks or Kernel variants of classic fault detection techniques such as Kernel Principal Component Analysis or Kernel Partial Least Squares can be used. When the relations are linear, and the probability density function of the data follows a Gaussian/Normal distribution, many techniques can be employed such as Principal Component Analysis (PCA) and Partial Least Squares (PLS) [75], Canonical Variate Analysis (CVA) [69], Gaussian Mixture Models (GMM) [79], and Hidden Markov Models (HMM) [52]. Conversely, if the probability density function does not follow a Gaussian distribution, then the fault detection model should be based on the method Independent Component Analysis (ICA) [3]. Some other techniques less commonly used for fault detection during steady modes are: Locality Preserving Projection [71], Factor Analysis [87], Neighborhood Preserving Embedding [55], and Local Outlier Factor [40, 57].

The fault detection methods which can be used for transitions depend on the data available for each transition, in other words, the frequency of occurrence of each transition. When the transition data has been recorded many times (high fre-

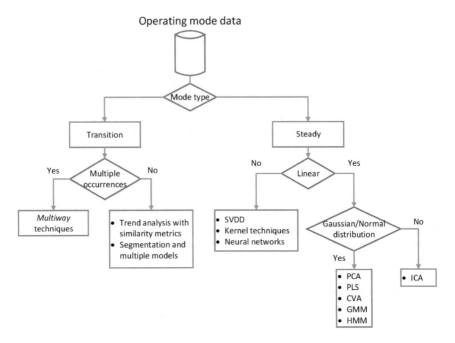

Fig. 4.3 Selection guide for fault detection strategy

quency) the *multiway* extensions of the classic multivariate statistical methods can be
used under the assumption of the repeatability of the transition pattern [12, 13, 16,
33, 81]. The repeatability condition implies that the trajectories of the variables
form a pattern that is repeatable within certain variability bounds. In most process
industries this condition is satisfied since control systems are used to manipulate the
predefined operating modes. The main problem to be tackled when using *multiway*
methods is the transition alignment problem: how to handle instances of the same
transition that present a different duration.

Many continuous processes are designed to operate mostly under steady mode
conditions and the transitions do not occur frequently. Two approaches to fault detection during transitions, in this case, are trend analysis and multiple fault detection
models. Trend analysis relies on the description of the transition based on a dictionary of individual shapes (often called *primitives*) and time-series similarity measures such as *Dynamic Time Warping* [58–60]. This approach requires a considerable
design effort as well as a high computational load to run the algorithms. Therefore,
a more suitable approach consists of using multiple fault detection models for each
single transition. The transition is divided into sequential time segments and a model
is built for each one [25, 26, 49, 63]. Each local fault detection model is used sequentially based on the evolution of the transition. The number of sub-models depend on
how the transition is segmented.

In the next sections, methods for fault detection of steady modes and transitions are presented in more detail.

4.3.1 Principal Component Analysis

Principal Component Analysis (PCA) is one of the most widely used data-based techniques for the diagnosis of industrial systems. It has been applied to reduce dimensionality, and is optimal in terms of capturing variance in the data [2, 44]. This representation of the data in a smaller space can improve the ability to detect and isolate faults. In addition, it can be useful in identifying the variables responsible for the fault or / and the variables most affected by the fault.

The original data space is transformed by maximizing the information in terms of the attribute's variability. The generative model obtained allows to capture the linear relationships among the physical variables [6]. The starting point of PCA and every monitoring method is a data set $X \in \Re^{n_n \times m}$ of m variables and n_n observations belonging to a single mode. Given the covariance matrix of the data $\Sigma = (X'X)/(n_n - 1)$, the eigenvectors associated with the a most significant eigenvalues of Σ form the matrix $P \in R^{m \times a}$ such that $a < m$. Variables should be scaled to zero mean and unit variance before applying the PCA transformation due to their different units. The statistical model obtained through PCA can be mathematically represented as follows

$$X = \tilde{X} + E = TP' + E \qquad (4.1)$$

where the principal component subspace ($\tilde{X}$) accounts for the main process variability and the residual subspace $E \in R^{n_n \times m}$ accounts for low variability usually associated with noise [48]. Thus, the residual part roughly represents the equation errors in mathematical models built from physical laws. Besides the criteria of selecting the number of components according to the eigenvalues significance, other criteria can be used, e.g. the Scree test and cross-validation [6].

Process monitoring is performed by using two indexes that quantify the model's mismatch with respect to the process behavior. The T^2 statistic index measures the signal's deviation in the principal component subspace. The Q (or Squared Prediction Error (SPE)) statistic index measures the squared prediction error spanned by the reduced model. A threshold is defined for each index based on a sample data set of the process normal behavior. If the assumption of multivariate normal distribution of the data is met, the distribution of the statistics can be defined in advance, such that the thresholds can be directly calculated. Otherwise, Kernel Density Estimation must be employed to estimate the statistical distribution of the indexes and their respective thresholds. Furthermore, these thresholds can be heuristically fine-tuned based on a desired fault detection rate and/or a boundary false alarm rate [6].

The main assumptions considered when employing a PCA model for monitoring are: the process is linear, the variables are not correlated in time, the process parameters are not time-varying, and the multivariate data is unimodal and Gaussian-distributed [34, 48].

Since this method will be used for testing the multimode monitoring strategies in the case of steady modes, the Matlab© code on how to use it for fault detection is shown in Appendices E.1 and E.2.

4.3.2 Partial Least Squares

Partial Least Squares (PLS) also known as Projection to Latent Structures is a statistical technique that creates a linear regression model of the relationships among two sets of variables. A historical data set of n_n observations is formed by a group of process $X \in \Re^{n_n \times m}$ and quality variables $Y \in \Re^{n_n \times p}$, considering m inputs and p outputs, respectively. The PLS latent variable model can be mathematically expressed as follows

$$X = \tilde{X} + E = T P' + E \tag{4.2}$$

$$Y = \tilde{Y} + F = T Q' + F \tag{4.3}$$

where $E \in R^{n \times m}$ and $F \in R^{n \times p}$ represent the residual subspace of the input and output variables respectively, $P \in R^{m \times l}$ and $Q \in R^{p \times l}$ are called loading matrices and they are obtained by maximizing the covariance between the deflated input data and the output data. Hence, dimension reduction is first applied to extract $\mathbf{t} \in \Re^l$ most relevant latent variables ($l \le m$) which allow to reconstruct the outputs. The different algorithms to create the PLS model are described in [36]. The number of latent variables is usually adjusted through cross-validation to maximize the prediction power. It is suggested to scale the data matrices to zero mean and unit variance.

Process monitoring is performed by using two indexes that measure the model error with respect to the behavior of the output and input variables [36]. The thresholds for both indexes are calculated based on their statistical distributions in a similar fashion as PCA monitoring indexes. The main assumptions of the PLS model for process monitoring are the same as the PCA model.

Since the cost of stopping plant operations are high, the main interest in the application of PLS for process monitoring is to detect faults that may lead to a production quality problem. This matter is of special consideration for the chemical industry because quality variables are not generally measured with a high sampling frequency.

4.3.3 Independent Component Analysis

Independent Component Analysis (ICA) is a feature extraction technique that transforms the original variables through the maximization of the statistical independence among the projected variables in the new feature space [28]. It is well known that many of the variables monitored in a process are not independent. These measured variables can be combinations of independent variables that are not directly measured, which are assumed to be non-Gaussian and mutually independent. ICA has the ability to extract these factors or components that govern the behavior of the system.

Given a data set $X \in \mathfrak{R}^{n_n \times m}$ with m variables and n_n observations, the statistical model obtained through the ICA transformation can be mathematically represented as follows

$$X = \hat{X} + E = SA' + E \qquad (4.4)$$

where $S \in R^{n_n \times d}$ is formed by $d \leq m$ feature vectors called independent components, $A \in R^{m \times d}$ is called mixing matrix and $E \in R^{n_n \times m}$ represents the residual subspace. Most ICA algorithms involve the maximization of a non-Gaussianity measure to extract the independent components, i.e. the FastICA algorithm. The number of independent components is usually estimated by using the Scree test [74].

Process monitoring is performed by using three indexes. Two of them, monitor the model error regarding the behavior of the variables in the independent component subspace (I and I_e statistics can be used [72]). The third one quantifies the squared prediction error spanned by the statistical model (Q or SPE statistic). The process monitoring thresholds for these indexes are calculated via kernel density estimation because no specific distribution for the variables can be assumed a priori. The ICA-based model assumes that the process variables have non-Gaussian distributions excepting one of them [28]. Other assumptions of ICA monitoring model are: the variables are not correlated in time, the relationships among the variables are linear, and the process parameters are not time-varying.

It is necessary to emphasize that the objectives of PCA and ICA are not the same. PCA is a reduction technique that projects the correlated variables to a smaller set of new variables that are not correlated, and that retain the greatest amount of information from the original variables. However, its objective is only to de-correlate the variables and not make them independent. PCA can only impose independence up to second-order statistical information mean and variance, while restricting the direction of the vectors so that they will be orthogonal. ICA has no orthogonality restrictions and involves higher order statistics; that is, it not only de-correlates the data but also reduces the higher order statistical dependence. Therefore, independent components reveal more useful information than the principal components [5, 27]. This method will be then compared with PCA for testing the multimode monitoring strategies in the case of steady modes. The Matlab© code on how to use it for fault detection is shown in Appendices E.3 and E.4.

4.3.4 Support Vector Data Description

Support Vector Data Description (SVDD) is a variant of Support Vector Machines that defines a boundary for the distribution of a data set. Then, SVDD has been used for anomaly detection and one-class classification problems [18]. A boundary for the nominal data distribution is estimated by minimizing the volume of a hypersphere in a feature space F such that, it holds all observations leaving out outliers and noise data [61].

SVDD technique starts by transforming the original variable space onto a feature space through a non-linear transformation, $\Phi_{svdd} : \mathbf{x} \rightarrow F$. The following convex optimization problem is then solved to estimate the boundary of the nominal feature space distribution

$$\min_{R,b,\xi}\{R^2 + C \sum_{t=1}^{N} \xi\} \; such \; that \; \parallel f_{svdd}(\mathbf{x}) - b \parallel \leq R^2 + \xi \qquad (4.5)$$

where b and R are the center and the radius vectors of the hypersphere, respectively, C denotes the trade-off between the volume of the hypersphere and number of errors, f_{svdd} represents the transformation function (for example the Gaussian kernel function with bandwidth h can be used) which projects the data into F, and ξ indicates the slack terms which represent the probability that some of outliers are erroneously included into the boundary.

Process monitoring with SVDD-based model is based on the distance between the projection of the new observation $\Phi_{svdd}(\mathbf{x})$ and the center of the hypersphere b. If such distance is larger than the radius R a fault/non-nominal behavior is detected. The SVDD technique makes no assumption regarding the distribution of the variables allowing to define a nominal boundary even for the distribution of non-linearly related variables. Moreover, the nominal boundary defined through the SVDD technique is robust against outliers. A disadvantage to be highlighted is that the interpretation of the SVDD-based monitoring strategy is not as straightforward as PCA or PLS-based statistical monitoring models. Therefore, the analysis of the process behavior is not easy because the model has not a clear physical interpretation. The assumptions made when using SVDD for process monitoring are that the variables are not correlated in time and the process parameters are not time-varying. Nonetheless, this method will be compared with PCA and ICA for testing the multimode monitoring strategies in the case of steady modes. The Matlab© code on how to use it for fault detection is shown in Appendices E.5 and E.6. The Data Description toolbox is used to compute the support vectors [90].

4.3.5 Gaussian Mixture Models

The distribution of a data set can be modeled through a Gaussian Mixture Model (GMM). The GMM technique describes the data set through a finite number of local Gaussian distributions called Gaussian components. A GMM is described in compact notation as follows

$$\Theta = \{\{w_1, \mu_1, \Sigma_1\}, \{w_2, \mu_2, \Sigma_2\}, ..., \{w_K, \mu_K, \Sigma_K\}\} \tag{4.6}$$

where w_i, μ_i and Σ_i are the prior probability, mean vector and covariance matrix of the local Gaussian i respectively. The total number of distributions to be characterized is K. The parameters of the GMM can be estimated with the Expectation-Maximization algorithm, or the F-J algorithm [78].

The process monitoring indexes of GMM are formulated through probability theory. The posterior probability of an observation to belong to each Gaussian distribution is computed through Bayesian inference. The underlying assumption of GMM is that any arbitrary distribution can be described through a set of Gaussian distributions. The number of components, however, must be carefully chosen to avoid overfitting.

While the PCA, PLS or ICA-based monitoring methods allow to characterize unimodal distributions, GMM allows to describe multimodal data. Thus, GMM has been used directly for monitoring processes with multiple modes. In fact, the number of distributions can be associated with the number of steady modes such that the linear physical relationships among the variables in each mode are captured by Σ_i as well as the nominal steady operating point by μ_i. The parameter w_i can be interpreted as the probability of occurrence for mode i which is usually dictated by how much time the process operates under that mode. GMM can be combined with other techniques to characterize non-linear and non-Gaussian distributions [3, 78]. Still, GMM are not recommended to monitoring transition modes because there is a high risk of overfitting if a transition mode is represented by multiple Gaussian components. Additionally, the GMM-based monitoring approach does not considers explicitly the correlations in time of the variables or time-varying process parameters.

4.3.6 Hidden Markov Model

A Hidden Markov model (HMM) describes the observation probability distribution of a data set together with the state transition probability distribution. HMM considers that the process generating the data set is represented by a finite number of states $\mathbf{S} = \{S_1, S_2, ..., S_k\}$ [50]. The transition probability among states as well as the probability that an observation belongs to a certain state can be estimated. The compact notation $\lambda = \{A_h, B_h, \pi\}$ describes a HMM as follows

- the probability transfer matrix

$$A_h = \{a_{ij}\}, \ 1 \le i, j \le k \tag{4.7}$$

where $a_{ij} = P(q_{t+1} = S_j | q_t = S_i)$ and q_t represents the hidden state at time t.
- the observation probability matrix

$$B_h = \{b_i(\mathbf{x}(t))\} \tag{4.8}$$

where $\mathbf{x}(t)$ is an observation at time t and $b_i(\mathbf{x}(t)) = P(\mathbf{x}(t)|q_t = S_i)$.
- the initial state probability distribution

$$\pi = \{\pi_i\} \tag{4.9}$$

where $\pi_i = P(q_1 = S_i), \sum_{i=1}^{k} \pi_i = 1$.

The parameters of a HMM are estimated though the Baum-Welch algorithm [22]. The distribution of each state is usually assumed to be a Gaussian Mixture [66]. A HMM allows to directly model multimode processes because it characterizes data set through multiple Gaussian distributions. More specifically, each state is associated with an operating mode. Mode identification to describe the multimode data is made by using the Viterbi algorithm [1]. Process monitoring is then achieved by assessing how the behavior of the process matches the HMM states through quantization measures [63, 77]. A HMM does not explicitly consider that the variables are correlated in time, and it assumes that the process parameters are not time-varying.

4.3.7   Transition Monitoring Model

The transition monitoring model detailed consists of using multiple fault detection models. The first step is to divide the transition into time segments and a model is built for each one. Once the start of a transition is identified, the first model is activated and the rest of the models will be used in a sequential fashion. The fault detection model to be used in each sub-segment is based on a distance evaluation approach.

When no transition has been identified and the mode is unknown then the similarity between the observation and the first segment of any of the transitions is evaluated. To build the model of each segment, the data set $X_{st} \in \Re^{n \times m}$ formed by m variables and n_{st} observations will be used. A threshold based on the Mahalanobis distance is defined as

$$distth_{st}(X_{st}) = max\{dist(\mathbf{x}_j, X_{st}) \ : \ \forall \mathbf{x}_j \in X_{st}\} \tag{4.10}$$

and

$$dist(\mathbf{x}_j, X_{st}) = (\mathbf{x}_j - \mu_{st})\Sigma_{st}^{-1}(\mathbf{x}_j - \mu_{st})^T \tag{4.11}$$

where μ_{st} is the mean vector and Σ_{st} the covariance matrix of X_{st}. For observation i, $dist_i(\mathbf{x}_i, X_{st})$ is calculated and considered normal if the following condition is satisfied

$$Dl = \frac{dist_i}{dlim_{st}} \leq \alpha \tag{4.12}$$

where the parameter $0 \leq \alpha \leq 1$ represents the confidence threshold for fault detection. The confidence threshold can be selected according to the probability density distribution of the statistic Dl. Considering that $0 \leq Dl \leq 1$ and assuming a Gaussian distribution for Dl then a fault detection threshold of $\alpha = 0.95$ should guarantee that a maximum of 5% of the observations of the transition X_{st} are considered as false positives. However, to avoid this assumption the distribution of Dl can be estimated through Kernel density estimation. Therefore, α is tuned to control the number of false alarms based on the normal behavior of the process.

Once a transition starts, the fault detection is evaluated based on the consecutive segments until a new steady mode is reached. This approach is appealing because multiple instances of the same transition are not required.

4.4 Cases of Study

In this section, the operating mode identification and fault detection problem is solved for the three cases of study. The fault detection method which will be applied for steady mode monitoring is the most used one throughout the literature: principal component analysis [68]. It will be illustrated how to calibrate the parameters of this method through the procedure presented at the beginning of this chapter. The transition monitoring model described in this chapter will be used for fault detection during transitions.

4.4.1 Continuous Stirred Tank Heater

The number of steady modes and transitions for this process were identified in the previous chapter. The dynamic response of this process is fast (in the order of seconds) compared to other type of systems. Therefore, the monitoring of transitions could be ignored without a great impact on the number of false alarms. Two mode identification/fault detection schemes will be tested for this case study: the mode selection function and the Bayesian fusion approach. In both cases, only data of the normal behavior of the process will be used to develop the fault detection models. This approach obeys to the fact that little data of faults is available in most industries.

The PCA method is selected for steady mode modeling and fault detection. The following parameters must be calibrated for each steady mode:

- Number of principal components. Range: [1–3];
- Confidence threshold for the detection statistics. Range: [0.95–0.99].

To calibrate the parameters of each steady mode a grid search can be developed. The number of false alarms (FAR evaluation measure) is the performance measure to be considered for adjusting the parameters. In general, the FAR should not be higher than 5%. This means that considering five in one hundred observations as false alarms is acceptable and it will not harm the trust of the process operators in the software that runs the fault diagnosis application.

The range of principal components for the grid search goes from one to three. Assuming that the variables follow a multivariate Gaussian distribution, the minimum theoretical value to be considered for the confidence threshold of the detection statistics (T^2 and SPE for the PCA model) should be 0.95. Figure 4.4 shows the FAR (%) obtained for the range of parameters. Although the confidence threshold of 0.95 should guarantee a FAR lower than 5% the FAR achieves values as high as 9.5%. The reason for this result is that the data distribution does not follow a multivariate Gaussian distribution. For this process in particular, a confidence threshold larger than 0.97 is required to ensure a FAR lower than 5% independently of the number of principal components retained. If the parameters corresponding to the minimum FAR are selected there is a risk of overfitting the data set. This could cause a loss of sensitivity for the posterior fault detection task. Therefore, a good trade-off for selecting the parameters is a confidence threshold of 0.99 and two components retained.

Figure 4.5 shows the fault detection statistics for a period of time from normal behavior. Most false alarms occur during transitions. The overall FAR for the entire period of study is 3.9% so the FAR is maintained lower than 5%. This demonstrates empirically that the false alarms that occur during transitions do not have an important effect in the overall performance. The reason behind this result is that transitions last a very short period of time compared to the duration of steady modes.

Figures 4.6 and 4.7 show the monitoring results for the mode selection function approach. The evaluation measures are a FAR of 3.62% and a FDR 97.64%. The performance measures show that a reasonable performance can be achieved when using this monitoring approach. The false alarms are mainly generated during mode changes where the transitions occur. However, the number of false alarms is not significant compared entire operation of the process under normal conditions.

Figures 4.8 and 4.9 show the monitoring results for the Bayesian fusion approach. The evaluation measures are a FAR of 3.79% and a FDR 97.66%. The performance measures show a similar performance compared to the mode selection approach. Again, the false alarms are mainly generated during mode changes where the transitions occur. The two mode identification/fault detection schemes have a similar performance because for this process the transitions present a short duration so only monitoring models for the steady modes are built. The successful results achieved by applying a simple technique as PCA owes to the approximate linear relationship among the variables of the process.

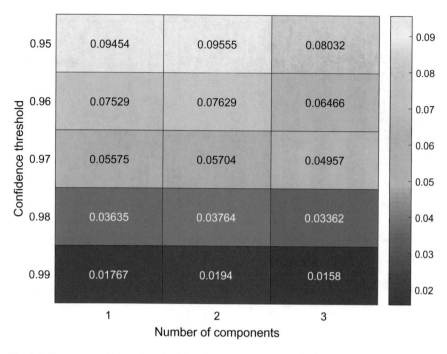

Fig. 4.4 Parameter grid based on the false alarm rate (%) for the PCA models built with mode 1 of the CSTH process

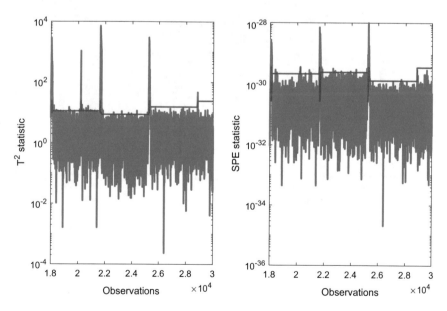

Fig. 4.5 Fault detection statistics for the normal process behavior of the CSTH process

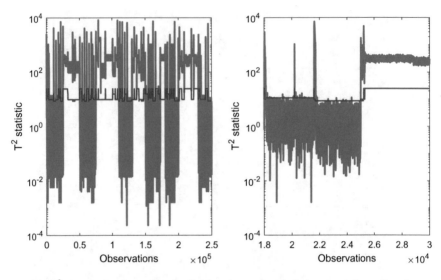

Fig. 4.6 T^2 statistic for the mode selection function approach during the entire testing data set (left) and zoom in an interval (right)

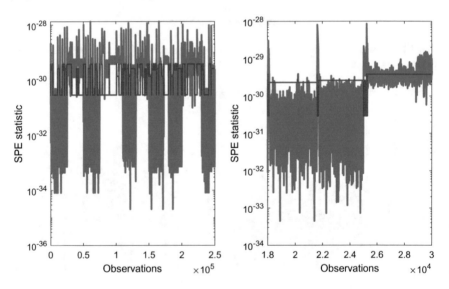

Fig. 4.7 SPE statistic for the mode selection function approach during the entire testing data set (left) and zoom in an interval (right)

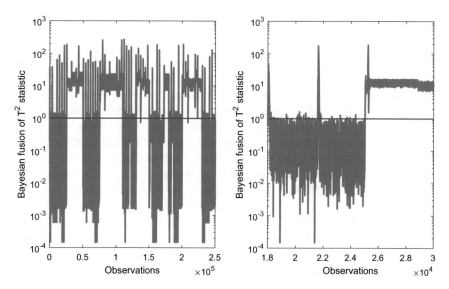

Fig. 4.8 Bayesian fusion of T^2 statistic during the entire testing data set (left) and zoom in an interval (right)

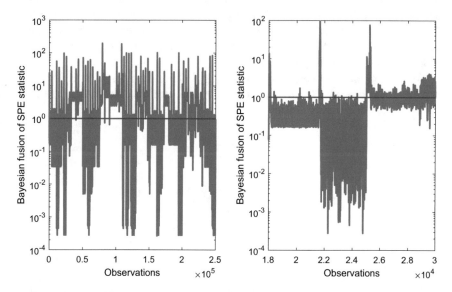

Fig. 4.9 Bayesian fusion of SPE statistic during the entire testing data set (left) and zoom in an interval (right)

4.4.2 Hanoi Water Distribution Network

The main challenge for this case study is to distinguish between faults and normal data in the testing data set. The mode identification and fault detection for the HWDN benchmark depend on both the time of the day and variations in the consumption patterns from day to day. In the previous chapter, the normal data set was divided into observations per sample time and clustering was performed for each subgroup. This data division removes the high non-stationary behavior of the demand throughout the day. Clustering allowed to identify the number of patterns in the day to day variability. The clusters identified represent steady modes. Therefore, any of the three mode identification/ fault detection schemes can be applied in this case. The data transformation approach will be tested for mode identification in this case study because the number of models created is smaller than those of the other approaches. Because the sampling time for this benchmark is 30 min a total of 48 mode identification/fault detection models will be developed as it is described in Fig. 4.10. If the other approaches would be used, then the number of models for each sample time depends on the number of clusters identified.

To develop the fault detection models for each sample time the Z-score scaling technique is applied first to each observation according to the mean and standard deviation of the cluster it belongs. This implies that the multimodal feature is removed for each sampling time. The result of this transformation for two variables measured is shown in Fig. 4.11. Afterwards, a fault detection model based on the PCA method is developed for each sampling time.

To develop the fault detection models only data of the normal behavior of the process is considered. The main reason is that for most industries there are many

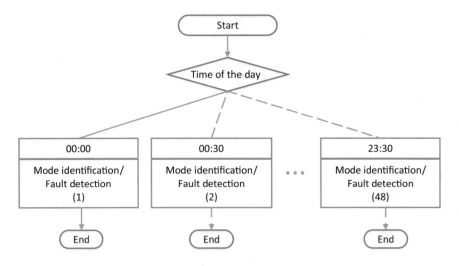

Fig. 4.10 Selection of mode identification/fault detection scheme according to the time of the day

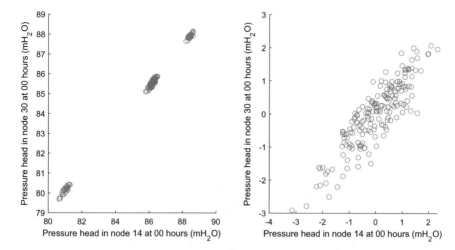

Fig. 4.11 Data transformation to remove the multimode feature. Original data (left) and data transformed (right)

data sets representing the typical behavior of the process but little data of faults. The two parameters to be set based on normal data for each model are

- Number of principal components. Range: [1–3];
- Confidence threshold for the detection statistics. Range: [0.95–0.99].

A grid search approach can be developed for searching the parameters of each PCA model. There is a total of 48 PCA models where each one is developed for a specific hour of the day (sampling time of 30 min with 24 h). The performance measure to be considered for adjusting the parameters is the number of false alarms. The same parameters will be used for all the models by assuming that the data distribution is similar. The FAR evaluation measure should be bounded/controlled below 5% as it was explained in the previous case study.

The range of possible values for the grid search must be established. The number of principal components goes from one to three. Assuming that the variables follow a multivariate Gaussian distribution, the minimum of the confidence threshold is 0.95. Fig. 4.12 shows the FAR (%) obtained for the range of parameters. Despite the theoretical confidence threshold of 0.95 should guarantee a FAR lower than 5% this is not meet for this process. The reason for this result is that the data distribution might not perfectly fit a multivariate Gaussian distribution. For this process in particular, a confidence threshold larger than 0.97 is required to ensure a FAR lower than 5% independently of the number of principal components retained. If the parameters corresponding to the minimum FAR are selected there is a risk of overfitting the data set. This could cause a loss of sensitivity for the posterior fault detection task. Therefore, a good trade-off for selecting the parameters is a confidence threshold of 0.99 and two components retained.

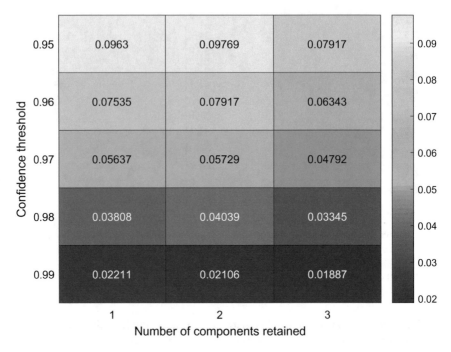

Fig. 4.12 Parameter grid based on the false alarm rate (%) for the PCA models built with the Hanoi data

Once the models are developed the performance of this approach for fault detection is evaluated. The results for the 6 data sets are shown in Fig. 4.13. The FAR remains below the 5% threshold thus the model calibration previously made is effective. As it is observed, the fault detection performance improves as the leak size increases. The nominal consumption in the nodes of this network are relatively high with a minimum, average and maximum of 16.67, 178.61 and 375 lps respectively. This means that the size of the leakages generated (varying between 25 and 75 lps) represents an average between 14 and 42% of nominal consumption of the nodes. This high variability in the consumption of the nodes also implies that the leakages that occur in nodes with small demand will probably be masked by the demand variability so they are not detectable. This explains why the FDR achieved for leakages with large size is 90%.

4.4.3   Tennessee Eastman Process

The multimode behavior of the TEP involves steady modes and transitions. The fault diagnosis strategies for each type of mode are different as it has been explained throughout this chapter. The appropriate mode identification approach is the mode selection function because of two reasons. One, there is no data transformation

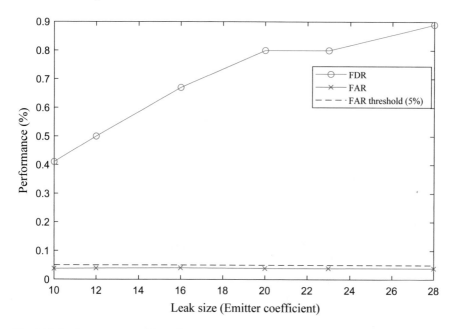

Fig. 4.13 Performance results for different leak sizes with the data transformation/fault detection approach based on PCA

scheme that allows to remove the multimode feature when modes of different type occur in the process. Two, it is complicated to develop a Bayesian fusion strategy that allows to combine the fault detection indices of many different mode types.

The mode identification/fault detection approach is depicted in Fig. 4.14. The main assumption of this approach is that between steady modes there always occur a transition. This assumption is satisfied for most continuous chemical processes with slow dynamics such as the TEP. The mode selection function to be used in this case is the model fitness that depends on the type of mode. It is assumed that the initial type of mode of the process is known. The first step is then to select the steady mode or transition that best fits the behavior of the process according to the first observation in $t = 1$. If the steady mode i is selected from the total I steady modes then the fitness of it's respective model is verified. If the behavior of the process does not fits the steady mode then it is assumed that a transition or a fault is occurring. Thus, the transition that best fits the current behavior is selected. If the behavior of the process fits the transition then this transition is selected as the current operating mode. Otherwise, it is considered that a fault is occurring. The same analysis is made when a transition is identified.

The PCA method and the transition monitoring approach presented in Sect. 4.3.7 will be used for modeling the steady modes and transitions, respectively. Therefore, the parameters of the PCA model for each steady mode will be first calibrated based on the same strategy explained in the previous Sect. 4.4.2. The performance measure

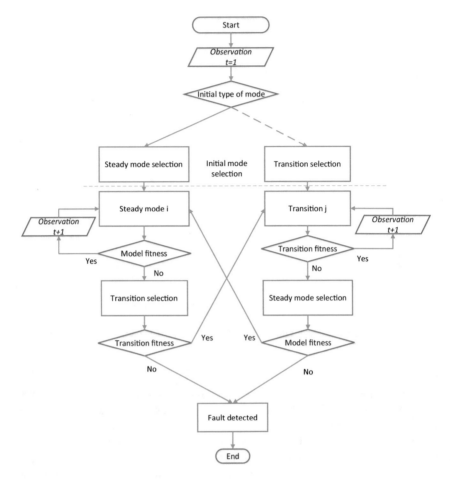

Fig. 4.14 Mode identification/fault detection scheme for processes with steady modes and transitions

used again is the FAR. Figure 4.15 shows the FAR (%) obtained for the range of parameters by considering mode 3. Similar to the situation of the Hanoi WDN, the theoretical confidence threshold of 0.95 does not guarantee a FAR less than 5%. To avoid overfitting a good compromise is to select a threshold of 0.98 with 15 principal components. The same procedure is used to calibrate the parameters of the other PCA models corresponding to the other steady modes identified.

The fault detection during transitions depends on two parameters: the window size and the confidence threshold for each transition. The ranges considered for the parameters are

- Window size. Range: [100–400];
- Confidence threshold for the detection statistic Dl. Range: [0.95–0.99].

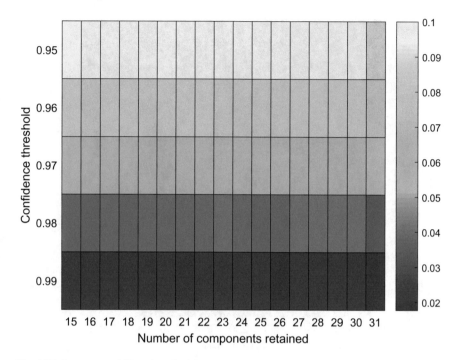

Fig. 4.15 Parameter grid based on the false alarm rate (%) for the PCA model of a steady mode of the TEP

Figure 4.16 shows the FAR (%) obtained for the range of parameters in the first transition (mode 3 to mode 1). In this case, the worst FAR equals to 2% for a window size of 100 and a confidence threshold of 0.95. Conversely, there are other set of parameters that guarantee a FAR as low as 0.004%. However, as it has been explained previously, the parameters should be selected to avoid overfitting. Thus, the parameters corresponding to a FAR of 2% are selected for this transition. Similarly, the rest of the parameters are selected for the other two transitions.

The fault detection statistics for the entire period of normal behavior are shown in Fig. 4.17a and b with logarithmic scale for the y axis. The fault detection statistics for the PCA model of each steady mode are T^2 and SPE. The statistic for the transition is Dl. For visualization purposes, each figure alternates between the steady mode statistics and the transition statistic. This approach allows to distinguish the mode changes graphically. As it is shown, the multiple model with decision function approach allows to track the mode changes successfully. The FAR for the entire period of study is 1.6% so the FAR is maintained lower than 5%. This demonstrates empirically that by adjusting the parameters of each fault detection model individually allows to guarantee the overall performance.

The fault detection results for the 6 cases described previously are shown in Table 4.1. The fault detection performance varies depending on the type of fault,

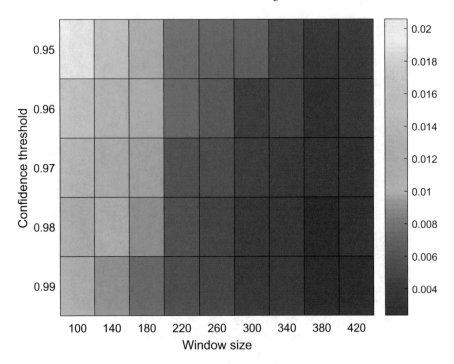

Fig. 4.16 Parameter grid based on the false alarm rate (%) for the transition monitoring model of a transition of the TEP

and the operating mode where it occurs. All faults are detected more easily on average during the transitions than during the steady modes. This occurs because the minimum deviation of the predefined set-points during transitions causes problems to the control systems of this non-linear process. In some cases, the control systems are able to compensate the effect of faults during steady modes making them more difficult to detect.

The first fault scenario corresponds to a step fault that affects the feed ratio of two chemical reactants of the process (named A and C). This fault has a strong impact in the process performance, thus it is detected with a high FDR independently of the mode where it occurs. The situation is different for fault scenarios 2, 4, 5 and 6. In all these scenarios the FDR performance depends on the operating mode.

For instance, the fault simulated in the second scenario (corresponding to fault No. 5 in the original description of the process [9]) is barely detected under operating mode 1. This fault consists of a step deviation in the condenser water inlet temperature. This variable is not directly measured and its variability has an impact in the product separator temperature. However, the set-point of the separator temperature is different for operating modes 1 and 3 (91.7 and 83.4 °C). Therefore, when the process operates in mode 1 the effect of fault No. 5 is not very significant as it is shown in Fig. 4.18. The deviation caused by the fault is larger with respect to the variance in mode 3 than in mode 1.

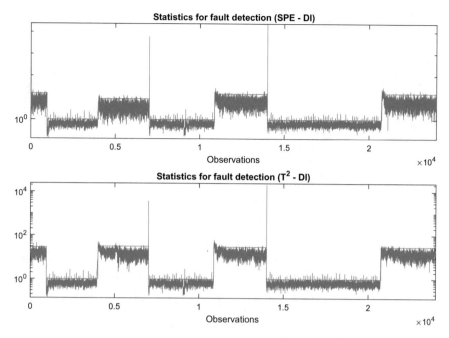

Fig. 4.17 Fault detection statistics for the normal process behavior of the TEP

Table 4.1 FDR (%) of faults for the TEP in different modes considering PCA for fault detection during steady modes

Fault scenario	Fault no	Mode 3	Mode 1	Transition
1	(1)	97.60	97.80	99
2	(5)	98.20	2	92.81
3	(10)	84.23	87.03	96.61
4	(13)	82.04	40.92	91.22
5	(14)	61.28	98.80	99.40
6	(16)	4.19	0	91.22
All	–	71.25	54.425	95.04

The third fault scenario (fault No. 10 in the original description of the process [9]) corresponds to a random variation in the feed temperature of reactant C. Similar to variable affected by fault No. 5, the variable corresponding to fault No. 10 is not measured. However, in this case, the effect of the fault in the stripper temperature makes it detectable in both modes. This can be observed in Fig. 4.19.

Finally, the sixth scenario (fault No. 16 in the original description of the process [9]) corresponds to an unknown fault. In this case, the fault detection performance for this fault is bad independently of the steady operating mode.

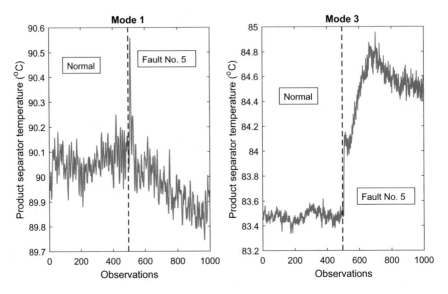

Fig. 4.18 Deviation caused by fault No. 5 in different operating modes of the TEP

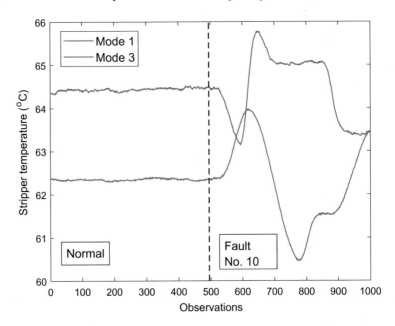

Fig. 4.19 Deviation caused by fault No. 5 in different operating modes of the TEP

4.4.3.1 Comparison with Different Steady State Monitoring Methods

In the previous experiments PCA was used to monitor the behavior of the system under steady state conditions. PCA takes into account the consistency the linear relationships among the system variables. However, there are other monitoring techniques that might improve the performance for this task. Specifically, Independent Component Analysis and Support Vector Data Description might offer some advantages over PCA for monitoring systems where the variables have a non-Gaussian distribution or the relation among them is non-linear, respectively.

The main parameters to be adjusted for ICA are the number of independent components to be retained and the confidence threshold for the fault detection statistics. ICA estimates the independent components by solving an optimization problem through a fixed-point algorithm. Therefore, there are other parameters that must be set in practice. The convergence threshold for the algorithm, the maximum number of iterations, and the non-Gaussianity measure are set to 0.01, 1000, and the Gauss function, respectively. If convergence is not achieved then the algorithm must be executed again. Also, the independent components are not estimated in any particular order. They must then be sorted once they have been estimated according to some criteria. In general, the amount of information in terms of variance is considered a widely accepted criteria [88]. The FastICA algorithm is used in this book to estimate the independent components [89].

A grid-search through the number of independent components and the confidence interval is performed. The results are similar to those of the PCA monitoring method with 15 components and a confidence a threshold of 0.98. The FAR obtained for the entire normal data is 2.07% thus the pre-defined theoretical FAR is achieved in this case. The results for the fault scenarios are shown in Table 4.2. The average FDR remains almost the same as the one obtained for PCA. Some improvement is achieved for fault 10. This fault is a random variation of one variable and the ICA model allows to detect this type of fault faster than PCA (fault detection delay of ICA is 21 min and fault detection delay of PCA is 30 min).

SVDD allows to model non-linear relationships among the variables. In addition, it tolerates the existence of outliers in the training data set. This is an advantage over PCA or ICA where techniques for cleaning the outliers must be applied before processing the data. The parameters to be adjusted for SVDD are the fraction of outliers allowed in the data set and the bandwidth of the Radial-Basis Kernel function used to map the observations to a hyper-sphere. Since the TEP data set does not contain outliers the fraction of outliers allowed is set a small value (~0.004). The bandwidth is then adjusted by considering the FAR as performance measure through k-fold cross validation on the training data set.

The bandwidth of the SVDD method is adjusted to remain below 5%. The FAR obtained for the entire normal data is 3% which can be considered as satisfactory. The results for the fault scenarios are shown in Table 4.3. The performance for faults occurring during mode 3 improves on average over the PCA and ICA methods. This is mainly the consequence of being able to detect the sixth fault scenario to some

Table 4.2 FDR (%) of faults for the TEP in different steady modes considering ICA for fault detection

Fault scenario	Fault no	Mode 3	Mode 1
1	(1)	97.60	98.60
2	(5)	98.20	1.4
3	(10)	84.21	90.82
4	(13)	82.83	43.51
5	(14)	61.3	98.60
6	(16)	3.19	1.4
All	–	71.22	55.72

Table 4.3 FDR (%) of faults for the TEP in different steady modes considering SVDD for fault detection

Fault scenario	Fault No	Mode 3	Mode 1
1	(1)	97.60	96.21
2	(5)	98.20	1.4
3	(10)	84.43	60.68
4	(13)	85.63	44.31
5	(14)	61.28	96.21
6	(16)	25.35	0
All	-	75.41	49.80

extent. However, for fault occurring during mode 1 the performance is worse than PCA and ICA by 5%. This means that the method is not able to properly characterize the data from that mode.

It should be highlighted that the FDR for some fault scenarios is the same regardless of the monitoring method. This is a consequence of the effect of the fault on the process dynamics. When a fault has a significant impact on the system all methods are able to detect it quickly thus all performance measures are equal or similar. This is the case for the step faults (scenarios 1 and 2) and the sticking valve fault (scenario 5).

Remarks

Monitoring of multimode continuous processes is a challenging task due to the existence of different types of modes: steady and transitions. Three operating mode identification strategies were presented in this chapter to cope with this problem. Fault detection methods to be used during steady modes and transitions were presented. The most used fault detection method was applied to three cases of study which have different features. It is essential to keep the false alarm rate low to maintain the reliability of the fault detection application. This is especially challenging during transitions. Moreover, whenever there are long transitions, the best mode identification scheme to be used is the one that incorporates multiple fault detection models and a function for model selection. This fact is illustrated in the TEP case of study where long transitions with low occurrence must be monitored.

References

1. Afzal, M.S., Tan, W., Chen, T.: Process monitoring for multimodal processes with mode-reachability constraints. IEEE Trans. Ind. Electron. **64**(5), 4325–4335 (2017)
2. Chaeng, H., Nikus, M., Jämsä Jounela, S.: Evaluation of pca methods with improved fault isolation capabilities on a paper machine simulator. Chemom. Intell. Lab. Syst. **92**(2), 186–199 (2008)
3. Chen, J., Yu, J.: Independent component analysis mixture model based dissimilarity method for performance monitoring of Non-Gaussian dynamic processes with shifting operating conditions. Ind. Eng. Chem. Res. **53**(13), 5055–5066 (2014)
4. Chen, J., Zhang, X., Zhang, N., Guo, K.: Fault detection for turbine engine disk using adaptive Gaussian mixture model. J. Syst. Control Eng. **231**(10), 827–835 (2017)
5. Cheng, C., Huang, K.: Applying ica monitoring and profile monitoring to statistical process control of manufacturing variability at multiple locations within the same unit. Int. J. Comput. Integr. Manuf **27**(11), 1055–1066 (2014)
6. Chiang, L.H., Russell, E.L., Braatz, R.D.: Fault Detection and Diagnosis in Industrial Systems. Springer-Verlag (2001)
7. Demšar, J.: Statistical comparisons of classifiers over multiple data sets. J. Mach. Learn. Res.**7**(Jan), 1–30 (2006)
8. Dougherty, G.: Pattern Recognition and Classification. Springer, New York, USA (2013)
9. Downs, J.J., Vogel, E.F.: A plant-wide industrial problem process. Comput. Chem. Eng. **17**(3), 245–255 (1993)
10. Du, W., Fan, Y., Zhang, Y.: Multimode process monitoring based on data-driven method. J. Frankl. Inst. **354**, 2613–2627 (2017)
11. Du, W., Tian, Y., Qian, F.: Monitoring for nonlinear multiple modes process based on LL-SVDD-MRDA. IEEE Trans. Autom. Sci. Eng. **11**(4), 1133–1148 (2014)
12. Duchesne, C., Kourti, T., MacGregor, J.E.: Multivariate monitoring of startups, restarts and grade transitions using projection methods. In: American Control Conference, pp. 5423–5426. IEEE, Denver, Colorado (2003)
13. Duchesne, C., Kourti, T., Macgregor, J.F.: Multivariate SPC for startups and grade transitions. AIChE J. **48**(12), 2890–2901 (2002)
14. Fazai, R., Taouali, O., Harkat, M.F., Bouguila, N.: A new fault detection method for nonlinear process monitoring. Int. J. Adv. Manuf. Technol. **87**(9–12), 3425–3436 (2016)
15. Feital, T., Kruger, U., Dutra, J., Pinto, J.C., Lima, E.L.: Modeling and performance monitoring of multivariate multimodal processes. AIChE J **59**(5), 1557–1569 (2013)
16. Garcia-Alvarez, D., Fuente, M.J., Sainz, G.I.: Fault detection and isolation in transient states using principal component analysis. J. Process Control **22**, 551–563 (2012)
17. Ge, Z., Song, Z.: Online monitoring of nonlinear multiple mode processes based on adaptive local model approach. Control Eng. Pract. **16**, 1427–1437 (2008)
18. Ge, Z., Song, Z., Gao, F.: Review of recent research on data-based process monitoring. Ind. Eng. Chem. Res. **52**, 3543–3562 (2013)
19. Ha, D., Ahmed, U., Pyun, H., Lee, C.j., Baek, K.H., Han, C.: Multi-mode operation of principal component analysis with k-nearest neighbor algorithm to monitor compressors for liquefied natural gas mixed refrigerant processes. Comput. Chem. Eng. **106**, 96–105 (2017)

20. Haghani, A., Jeinsch, T., Ding, S.X., Koschorrek, P., Kolewe, B.: A probabilistic approach for data-driven fault isolation in multimode processes. IFAC Proc. Vol. **47**(3), 8909–8914 (2014)

21. Haghani, A., Krueger, M., Jeinsch, T., Ding, S.X., Engel, P.: Data-driven multimode fault detection for wind energy conversion systems. IFAC-PapersOnLine **48**(21), 633–638 (2015)

22. Han, J., Kamber, M., Pei, J.: Data Mining Concepts and Techniques. Elsevier (2012)

23. He, Q.P., Wang, J.: Statistical process monitoring as a big data analytics tool for smart manufacturing. J. Process Control (2017)

24. He, X.B., Yang, Y.P.: Variable MWPCA for adaptive process monitoring. Ind. Eng. Chem. Res. **47**, 419–427 (2008)

25. He, Y., Ge, Z., Song, Z.: Adaptive monitoring for transition process using dynamic mutual information similarity analysis. In: Chinese Control and Decision Conference, pp. 5832–5837 (2016)

26. He, Y., Zhou, L., Ge, Z., Song, Z.: Dynamic mutual information similarity based transient process identification and fault detection. Can. J. Chem. Eng. (2017). https://doi.org/10.1002/cjce.23102

27. Hsu, C., Chen, M., Chen, L.: A novel process monitoring approach with dynamic independent component analysis. Control Eng. Pract. **18**, 242–253 (2010)

28. Hyvärinen, A., Karhunen, J., Oja, E.: Independent Component Analysis. John Wiley & Sons, Inc. (2001)

29. Jiang, Q., Huang, B., Yan, X.: GMM and optimal principal components-based Bayesian method for multimode fault diagnosis. Comput. Chem. Eng. **84**, 338–349 (2016)

30. Jin, H.D., Lee, Y.h., Lee, G., Han, C.: Robust recursive principal component analysis modeling for adaptive monitoring. Indust. Eng. Chem. Res.**45**, 696–703 (2006)

31. Kodamana, H., Raveendran, R., Huang, B.: Mixtures of probabilistic PCA with common structure latent bases for process monitoring. IEEE Trans. Control Syst. Technol. (2017). https://doi.org/10.1109/TCST.2017.2778691

32. Kourti, T.: Process analysis and abnormal situation detection: from theory to practice. IEEE Control Syst. Mag. **22**(5), 10–25 (2002)

33. Kourti, T.: Multivariate dynamic data modeling for analysis and statistical process control of batch processes, start-ups and grade transitions. J. Chemom. **17**(1), 93–109 (2003)

34. Kruger, U., Xie, L.: Statistical Monitoring of Complex Multivariate Processes. John Wiley & Sons, Inc. (2012)

35. Lee, Y.h., Jin, H.D., Han, C.: On-Line process state classification for adaptive monitoring. Indust. Eng. Chem. Res. **45**(9), 3095–3107 (2006)

36. Li, G., Qin, S.J., Zhou, D.: Geometric properties of partial least squares for process monitoring. Automatica **46**(1), 204–210 (2010)

37. Li, H., Wang, H., Fan, W.: Multimode process fault detection based on local density ratio-weighted support vector data description. Indust. Eng. Chem. Res. **56**(9), 2475–2491 (2017)

38. Liu, J.: Fault detection and classification for a process with multiple production grades. Indust. Eng. Chem. Res. **47**(21), 8250–8262 (2008)

39. Lou, Z., Wang, Y.: Multimode continuous processes monitoring based on hidden semi-markov model and principle component analysis. Indust. Eng. Chem. Res. **56**(46), 13800–13811 (2017)

40. Ma, H., Hu, Y., Shi, H.: Fault detection and identification based on the neighborhood standardized local outlier factor method. Indust. Eng. Chem. Res. **52**(6), 2389–2402 (2013)

41. Ma, L., Dong, J., Peng, K.: Root cause diagnosis of quality-related faults in industrial multimode processes using robust Gaussian mixture model and transfer entropy. Neurocomputing **285**, 60–73 (2018)

42. Ma, Y., Shi, H., Wang, M.: Adaptive local outlier probability for dynamic process monitoring. Chin. J. Chem. Eng. **22**, 820–827 (2014)

43. Ma, Y., Song, B., Shi, H., Yang, Y.: Neighborhood based global coordination for multimode process monitoring. Chemom. Intell. Lab. Syst. **139**, 84–96 (2014)

44. Monroy, I., Benitez, R., Escudero, G., Graells, M.: Enhanced plant fault diagnosis based on the characterization of transient stages. Comput. Chem. Eng. **37**(10), 200–213 (2012)

45. Natarajan, S., Srinivasan, R.: Multi-model based process condition monitoring of offshore oil and gas production process. Chem. Eng. Res. Des. **88**(5–6), 572–591 (2010)
46. Peng, X., Tang, Y., Du, W., Qian, F.: An online performance monitoring and modeling paradigm based on just-in-time learning and extreme learning machine for dynamic non-gaussian chemical process monitoring. Indust. Eng. Chem. Res. **56**(23), 6671–6684 (2017)
47. Peng, X., Tang, Y., Du, W., Qian, F.: Multimode process monitoring and fault detection: a sparse modeling and dictionary learning method. IEEE Trans. Indust. Electron. **64**(6), 4866–4875 (2017)
48. Qin, S.J.: Survey on data-driven industrial process monitoring and diagnosis. Annu. Rev. Control **36**, 220–234 (2012)
49. Quiñones-Grueiro, M., Prieto-Moreno, A., Llanes-Santiago, O.: Modeling and monitoring for transitions based on local kernel density estimation and process pattern construction. Indust. Eng. Chem. Res. **55**(3), 692–702 (2016)
50. Rabiner, L.R.: A tutorial on hidden Markov models and selected applications in speech recognition. Proc. IEEE **77**(2), 257–286 (1989)
51. Ralston, P., Depuy, G., Graham, J.: Graphical enhancement to support pca-based process monitoring and fault diagnosis. ISA Trans. **43**(4), 639–653 (2004)
52. Rashid, M.M., Yu, J.: Hidden Markov model based adaptive independent component analysis approach for complex chemical process monitoring and fault detection. Indus. Eng. Chem. Res. **51**(15), 5506–5514 (2012)
53. Ren, X., Tian, Y., Li, S.: Vine copula-based dependence description for multivariate multimode process monitoring. Indus. Eng. Chem. Res. **54**(41), 10001–10019 (2015)
54. Sammaknejad, N., Huang, B.: Operating condition diagnosis based on HMM with adaptive transition probabilities in presence of missing observations. AIChE J. **61**(2), 477–493 (2015)
55. Song, B., Ma, Y., Shi, H.: Multimode process monitoring using improved dynamic neighborhood preserving embedding. Chemom. Intell. Lab. Syst. **135**, 17–30 (2014)
56. Song, B., Shi, H.: Temporal-spatial global locality projections for multimode process monitoring. IEEE Access **6**, 9740–9749 (2018)
57. Song, B., Shi, H., Ma, Y., Wang, J.: Multisubspace principal component analysis with local outlier factor for multimode process monitoring. Indus. Eng. Chem. Res. **53**(42), 16453–16464 (2014)
58. Srinivasan, R., Qian, M.S.: Off-line temporal signal comparison using singular points augmented time warping. Indus. Eng. Chem. Res. **44**(13), 4697–4716 (2005)
59. Srinivasan, R., Qian, M.S.: Online fault diagnosis and state identification during process transitions using dynamic locus analysis. Chem. Eng. Sci. **61**(18), 6109–6132 (2006)
60. Srinivasan, R., Qian, M.S.: Online temporal signal comparison using singular points augmented time warping. Indus. Eng. Chem. Res. **46**(13), 4531–4548 (2007)
61. Tax, D.M., Duin, R.P.: Support vector data description. Mach. Learn. **54**(1), 45–66 (2004)
62. Wang, F., Tan, S., Shi, H.: Hidden Markov model-based approach for multimode process monitoring. Chemom. Intell. Lab. Syst. **148**, 51–59 (2015)
63. Wang, F., Tan, S., Yang, Y., Shi, H.: Hidden Markov model-based fault detection approach for multimode process. Indust. Eng. Chem. Res. **55**(16), 4613–4621 (2016)
64. Wang, F., Zhu, H., Tan, S., Shi, H.: Orthogonal nonnegative matrix factorization based local hidden Markov model for multimode process monitoring. Chin. J. Chem. Eng. **24**(7), 856–860 (2016)
65. Wang, G., Liu, J., Zhang, Y., Li, Y.: A novel multi-mode data processing method and its application in industrial process monitoring. J. Chemom. **29**(2), 126–138 (2014)
66. Wang, L., Yang, C., Sun, Y.: Multi-mode process monitoring approach based on moving window hidden Markov model. Indust. Eng. Chem. Res. **57**(1), 292–301 (2017)
67. Wang, X., Kruger, U., Irwin, G.W.: Process monitoring approach using fast moving window PCA. Indust. Eng. Chem. Res. **44**, 5691–5702 (2005)
68. Wang, Y., Si, Y., Huang, B., Lou, Z.: Survey on the theoretical research and engineering applications of multivariate statistics process monitoring algorithms: 2008–2017. Can. J. Chem. Eng. (2018). https://doi.org/10.1002/cjce.23249

69. Wen, Q., Ge, Z., Song, Z.: Multimode dynamic process monitoring based on mixture canonical variate analysis model. Indust. Eng. Chem. Res. **54**(5), 1605–1614 (2015)
70. Xie, X., Shi, H.: Dynamic multimode process modeling and monitoring using adaptive Gaussian mixture models. Indust. Eng. Chem. Res. **51**(15), 5497–5505 (2012)
71. Xie, X., Shi, H.: Multimode process monitoring based on Fuzzy C-means in locality preserving projection subspace. Chin. J. Chem. Eng. **20**(6), 1174–1179 (2012)
72. Xiong, Y., Bingham, D., Braun, W., Hu, X.: Moran's I statistic-based nonparametric test with spatio-temporal observations. J. Nonparametric Stat (2018). https://doi.org/10.1080/10485252. 2018.1550197
73. Xu, X., Xie, L., Wang, S.: Multimode process monitoring with PCA mixture model. Comput. Electr. Eng. **40**(7), 2101–2112 (2014)
74. Yin, S., Ding, S.X., Haghani, A., Hao, H., Zhang, P.: A comparison study of basic data-driven fault diagnosis and process monitoring methods on the benchmark Tennessee Eastman process. J. Process Control **22**(9), 1567–1581 (2012)
75. Yoon, S., Macgregor, J.F.: Fault diagnosis with multivariate statistical models part I: using steady state fault signatures. J. Process Control **11**, 387–400 (2001)
76. Yu, H.: A novel semiparametric hidden Markov model for process failure mode identification. IEEE Trans. Autom. Sci. Eng. **15**(2), 506–518 (2017)
77. Yu, J.: Hidden Markov models combining local and global information for nonlinear and multimodal process monitoring. J. Process Control **20**, 344–359 (2010)
78. Yu, J.: A nonlinear kernel Gaussian mixture model based inferential monitoring approach for fault detection and diagnosis of chemical processes. Chem. Eng. Sci. **68**(1), 506–519 (2012)
79. Yu, J., Qin, S.J.: Multimode process monitoring with Bayesian inference-based finite Gaussian mixture models. AIChE J. **54**(7), 1811–1829 (2008)
80. Zhang, S., Zhao, C.: Sationarity test and Bayesian monitoring strategy for fault detection in nonlinear multimode processes. Chemom. Intell. Lab. Syst. **168**, 45–61 (2017)
81. Zhang, Y., Dudzic, M., Vaculik, V.: Integrated monitoring solution to start-up and run-time operations for continuous casting. Annu. Rev. Control **27**(2), 141–149 (2003)
82. Zhao, C., Wang, W., Qin, Y., Gao, F.: Comprehensive subspace decomposition with analysis of between-mode relative changes for multimode process monitoring. Indust. Eng. Chem. Res. **54**(12), 3154–3166 (2015)
83. Zhao, F.z., Song, B., Shi, H.b.: Multi-mode process monitoring based on a novel weighted local standardization strategy and support vector data description. J. Cent. S. Univ. **23**(11), 2896–2905 (2016)
84. Zheng, J., Song, Z.: Linear subspace PCR model for quality estimation of nonlinear and multimode industrial processes. Indust. Eng. Chem. Res. **56**(21), 6275–6285 (2017)
85. Zheng, Y., Qin, S.J., Wang, F.l.: PLS-based Analysis for mode identification in multimode manufacturing processes. IFAC-PapersOnLine **48**(8), 777–782 (2015)
86. Zhou, L., Zheng, J., Ge, Z., Song, Z., Shan, S.: Multimode process monitoring based on switching autoregressive dynamic latent variable model. IEEE Trans. Indust. Electron. (2018). https://doi.org/10.1109/TIE.2018.2803727
87. Zhu, J., Ge, Z., Song, Z.: Recursive mixture factor analyzer for monitoring multimode time-variant industrial processes Recursive mixture factor analyzer for monitoring multimode time-variant industrial processes. Indust. Eng. Chem. Res. **55**(16), 4549–4561 (2016)
88. Lee, J., Yoo, C., Lee, I.: Statistical process monitoring with independent component analysis. J. Process Control **5**(14), 467–485 (2004)
89. Hyvarinen, A.: Fast and robust fixed-point algorithms for independent component analysis. IEEE Trans. Neural Netw. **10**(3), 626–634 (1999)
90. Tax, D.M.J.: DDtools, the Data Description Toolbox for Matlab. version 2.1.3 **Month**(Jan), (2018)

Chapter 5
Fault Classification with Data-Driven Methods

Abstract This chapter addressees the fault classification task with data-driven methods. Once a fault has been detected and the operating mode is identified, pattern recognition methods can be used to determine the cause of the fault. This is possible as long as there exists a data set representative of the fault to be classified. Four different classifiers are considered: neural networks (NN), support vector machines (SVD), maximum a posterior probability (MAP) and decision trees (DT). The procedure to solve the fault classification task is analyzed by using the three benchmarks used along this book.

5.1 Fault Classification

The fault classification task takes place after a fault is detected according to the data-driven monitoring loop presented in the previous chapter (Fig. 4.1). Fault classification comprises two sub-tasks: fault symptom identification and fault cause identification.

Solving the fault symptom identification task allows to identify the variables associated with the fault that has occurred. This information might be useful for process operators when systems have a high number of variables. This is because given the set of variables identified, the operators could be able to determine (based on their expert knowledge) which fault is occurring. However, the process of fault identification still depends on the experience of the operators to tackle the cause of the fault.

The fault cause identification allows to directly assess which fault is occurring. Thus, it saves time for the operators which may be critical to make decisions aiming at alleviating the problem. In order to determine the cause of the fault, it is necessary, for data-driven methods, to have some representative data of the fault behavior. There are two ways to obtain such data: (1) the fault has occurred before, and (2) the data is generated from some simulator of the process. This chapter focuses on the solution of this second sub-task.

© Springer Nature Switzerland AG 2021　　　　　　　　　　　　　　　　99
M. Quiñones-Grueiro et al., *Monitoring Multimode Continuous Processes*,
Studies in Systems, Decision and Control 309,
https://doi.org/10.1007/978-3-030-54738-7_5

5.1.1 Fault Classification as a Pattern Recognition Problem

Pattern recognition is a scientific discipline whose goal is the classification of *objects* into a number of categories or *classes* [23]. A class is interpreted as a set of *objects* that share some similarities (not necessarily identical) and which should be distinguishable from *objects* of other classes.

Different types of techniques are used for classifying *objects* depending on whether there exists some previous knowledge or not about the possible *classes*. If this knowledge is not available, then the pattern recognition task is concerned with the automatic discovery of regularities in data, mainly through the use of clustering techniques. On the other hand, if there are previous data available with *objects* from each category then the pattern recognition task consists of using the regularities to take actions such as classifying new *objects* [5]. Fault classification is concerned with the latter pattern recognition task.

In general, the pattern classification task can be defined as follows.

Definition 5.1 (*Pattern classification task*) Given a set of feature vectors $D_f\{\mathbf{x}\}_{j=1}^{n_f}$, $\mathbf{x}_j \in \Re^m$ and a set of categorical classes $\Omega = \{\omega_1, \omega_2, ..., \omega_Z\}$, the pattern classification task is defined as the task of finding a decision function: $g(D_f, \Omega): D_f \in \Re^{n \times m} \rightarrow \omega_z \in \Omega$ that associates the data set D_f with one of the Z classes.

One classic assumption of pattern recognition problems is that the classes are mutually exclusive [12]. As it can be seen from the definition above, the pattern classification task matches exactly with the fault cause identification task previously defined in Chap. 1.

It should also be noted that the fault cause recognition task is framed here as a supervised learning task from the perspective of machine learning techniques. Supervised learning is an inductive reasoning process, whereby a set of rules are learned from instances and a classifier algorithm is chosen or created that can apply these rules successfully to new instances [11].

5.1.2 Design of Fault Cause Identification Methods

The following procedure, based on the CRISP-DM methodology previously described in Chap. 1, can be applied to the design of fault cause identification methods.

1. The first step (business understanding) is related to the characterization of the process and data collection. If possible, a data set which is representative of the behavior of the process under all possible faults must be acquired. The variables to be collected must also be selected and some degree of expert knowledge about the process to be monitored is very useful to characterize in advance some features of the process (dynamic, non-linear, etc).

2. The second step, called data understanding, is related to the data-driven verification of the features of the process under the different fault conditions, as well as
   the quality of the collected data set. Additional properties of the process according
   to the data set could be identified in this step such as non-linearity, and dynamic
   behavior.
3. The third step involves the data preparation task. It involves data cleaning (outlier
   removal and noise filtering), handling missing data (possibly replacing a missing
   point by an estimate, and data synchronization (if variables have been acquired
   at different sampling rates)
4. The fourth step is modeling, and it is related to the selection and application of
   pre-processing techniques, pattern classification algorithms, and post-processing
   techniques. Pre-processing techniques may include variable scaling and normal-
   ization, feature extraction and feature selection. The selection and parameteriza-
   tion of the pattern classification algorithm is carried out through the process called
   training/testing, which will be described later. Post-processing techniques usually
   include a Bayesian analysis of the predictions made by the pattern classification
   algorithm, for making a final decision about the fault occurring in the system.
5. The fifth step is evaluation, and it is related to evaluate the performance of the
   fault cause identification based on the correct pattern classification. The confusion
   matrix and the accuracy are the most used evaluation measures.
6. The sixth step is related to the integration of the methods designed into a larger
   software system taking into account the hardware specifications.

5.2 Parameter Setting for Pattern Classification Algorithms

The fourth step of the design procedure previously described involves selecting and
adjusting the parameters of the classification method. Usually for each classification
method there is a set of parameters which are predefined, and a set of parameters
which are adjusted according to the characteristics of the data set. The former ones
will be called hyper-parameters in this book and the latter ones parameters. The
process of adjusting the parameters of a method involves evaluating the performance
on the classification task multiple times. The process of adjusting the parameters
according to the data set is called learning. Therefore, the hyper-parameters and the
parameters are distinguished by the fact that the latter ones are learned. Overall, the
following design elements must be considered:

- Evaluation function.
- Training/testing for parameter tuning.
- Hyper-parameter tuning.

5.2.1   Evaluation Function

The evaluation function determines the performance of the classification method. This function is usually called loss or cost function. This function evaluates mathematically the prediction error of the classification tool. The most popular one is the overall accuracy presented in Chap. 1 (Sect. 1.3.3). The overall accuracy should not be considered as a good performance measure when the number of observations of each class is not equal. This means that the classes are not balanced. In this case, it is a good approach to visualize the confusion matrix to determine the error for each class. There are other possible approaches to deal with the imbalanced data problem [9]. In the case of fault cause identification problem, it is always advised to visualize the confusion matrix independently of the overall accuracy obtained by the method.

5.2.2   Training/Testing for Parameter Tuning

The process of training means that the performance of the classification method is evaluated multiple times based on a function that evaluates its performance and its parameters are modified to improve that performance. This approach is usually called learning from examples because it involves the process of using data to determine the best parameters for the classifier. The best possible performance obtained at the end of this process is called training performance. Each classification algorithm uses a different optimization method to sequentially adjust its parameters based on the performance.

Once the best possible parameters of the classifier are obtained its performance is evaluated based on a data set which was not used for adjusting the parameters. The result is called testing performance and it determines how much the classifier has learned. The phenomena of having a classifier with a very good training performance but a very bad testing performance is called over-fitting. This means that the classifier is not able to generalize what has been learned during the training process.

The most popular strategy to avoid the over-fitting phenomena is called cross-validation [10]. For performing cross-validation, the data set that will be used for training is first divided into mutually exclusive and equal-sized subsets. Then, the classifier is trained with all subsets except one that is used for evaluating the testing performance. This is repeated leaving out a different subset each time and all performance measures obtained are averaged. This is known as leave-one-out cross-validation and the average of the performance measures will be a good approximation of how much the classifier has learned. The obtained evaluation is called cross-validation performance.

5.2.3 Hyper-parameter Tuning

The hyper-parameters of a classification method are the parameters which are defined before learning the parameters based on a data set. The strategy to set the hyper-parameters of a classifier is important since the overall performance may depend on their values. Manually setting the hyper-parameters of a classifier is an intensive task since there might be multiple possible combinations. Therefore, there are different strategies to automatically search the hyper-parameters of each method [3]. The main three search strategies are:

- Grid-search: This is the classic approach where the hyper-parameter space is manually defined in advance, and an exhaustive search is carried on. The range of each parameter is defined as well as the search interval inside this range. If there are some parameters with real-valued spaces, discretization is necessary before applying grid search. When the number of hyper-parameters is very large this strategy is very computationally demanding. However, the evaluation of each set can be run with distributed computing because generally each parameter set is independent of each other.
- Random-search: This approach selects the combinations of hyper-parameter randomly from the hyper-parameter space instead of performing an exhaustive enumeration of all possible combinations. The discretization process for real-valued parameters as well as the range definition is still necessary. This strategy can outperform Grid-search when only a small number of hyper-parameters affects the final performance of the algorithm [2].
- Optimization-based approaches: There are numerous optimization algorithms that can be used to adjust the hyper-parameters of classification methods. The most widely known are Bayesian optimization [21], and meta-heuristic optimization algorithms such as evolutionary optimization [15], and population-based optimization approaches [13]. Optimization approaches iteratively evaluate and update a promising hyper-parameter set trying to reveal the location of the optimum. They try to balance the search for hyper-parameters with an uncertain performance (exploration) and the search for hyper-parameters expected to be close to the optimum (exploitation).
  Bayesian optimization is a sequential approach for global optimization that does not require derivatives. Bayesian optimization develops a probabilistic model of a function mapping from hyper-parameter space to the performance on a validation data set. This model tries to capture the probability distribution over the objective function. The distribution function is usually modeled as Gaussian Processes [3]. Meta-heuristic optimization may use algorithms inspired by biological concepts such as evolution and population to adjust the hyper-parameters sequentially. Population-based optimization approaches learn both hyper-parameter and parameter values by continuously evaluating the performance of multiple hyper-parameter sets, discarding poorly performing models, and modifying the hyper-parameters from the better models. Meta-heuristic algorithms often perform well because they do not make any assumption about the underlying optimization landscape of the problem.

5.3 Pattern Classification Algorithms

There are numerous pattern classification algorithms which have been developed based on statistical techniques and artificial intelligence. The selection of the appropriate classifier depends on the characteristics of the process. However, there is no ultimate best classifier that works for any classification problem. This is the conclusion of the No Free Lunch Theorem [11]. Therefore, a good approach is to compare the performance of different classifiers to select the best one given the specific classification problem. In this Section, four classification algorithms of different nature are presented.

5.3.1 Bayes Classifier

The Bayes classifier is based on parameterizing a function $g_z(\mathbf{x})$ that models the probability density function for each class i. This implies that the probability of occurrence of a class ω_z (posterior probability) given the feature vector $\mathbf{x}$ is $g_z(\mathbf{x}) = P(\omega_z|\mathbf{x})$. Assuming that the probability distribution of each class is Gaussian, the (class-conditional) probability density function usually used to model $g_z(\mathbf{x})$ is the multivariate Gaussian [11]:

$$g_z(\mathbf{x}) = -\frac{(\mathbf{x} - \mu_\mathbf{z})' \Sigma_z^{-1}(\mathbf{x} - \mu_\mathbf{z})}{2} - \frac{ln|\Sigma_z|}{2} \qquad (5.1)$$

where μ_z and Σ_z are the mean vector and covariance matrix for class ω_z. When a covariance matrix is estimated for each class, and *a priori* probabilities are considered equal, the final decision function of the Bayesian classifier for a given observation becomes

$$g(\mathbf{x}, \Omega) = \max_{z=\omega_1,\omega_2,...,\omega_z} \{g_z(\mathbf{x})\} \qquad (5.2)$$

Usually $g_z(\mathbf{x})$ is called discriminant function, and the resulting decision surface that separates the different classes are quadratic. Therefore, the Bayes classifier is a quadratic classifier. Figure 5.1a shows a linear decision surface compared to Fig. 5.1b that shows a quadratic decision surface between the observations of two classes (red and blue).

The Bayes classifier is inspired in Bayesian decision theory, and tries to minimize the classification error probability based on the probability density function of each class [23]. This implies that this classifier is optimal (minimum classification error) when the classes are Gaussian distributed, and the distribution information is available [24]. One significant advantage of this classifier is that the only parameters to be estimated for each class are the mean vector and the covariance matrix, i.e.$\{\mu_z, \Sigma_z\}$. The Matlab$^©$ function for using this classifier is shown in Appendix F.1.

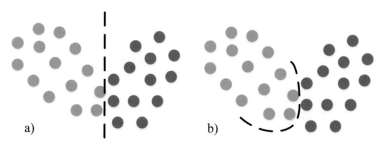

Fig. 5.1 Decision surfaces for the classification of two classes. **a** linear and **b** quadratic

5.3.2 Random Forests

Random forests is a classification technique based on the decision tree classifier. A decision tree is a recursive and partition-based classifier. The tree is formed by decision nodes which are interconnected such that each node represents a decision function

$$g(\mathbf{x}) = \mathbf{w}^T \mathbf{x} + b \qquad (5.3)$$

where $\mathbf{w}$ and b (bias) are used to define a hyperplane position; and $\mathbf{x} \in \Re^p$ denotes a feature vector. The value of $\mathbf{w}$ is restricted *a priori* to one of the standard basis vectors $\mathbf{e_1}, ..., \mathbf{e_p}$, where $\mathbf{e_1} \in \Re^p$ has a 1 for the j-th dimension and 0 for the others. This decision function specifies an axis-parallel hyperplane that splits the data space into regions.

A decision tree has a hierarchical structure that is formed based on the divide-and-conquer strategy. The tree starts with a root node which is considered the parent of the other nodes. Each node represents a decision function. Different branches are created starting from the root node and more nodes are created successively until a terminal node (called leaf) is reached, where the class is assigned.

An example of a classification problem with the respective decision tree is shown in Fig. 5.2a, b. There are three classes whose objects are represented spatially in Fig. 5.2a together with the respective decision surfaces. The root node is set for variable x and the leaf nodes are the classes/categories as shown in Fig. 5.2b.

A decision tree is created by selecting splitting points (variables) for which the decision function is calibrated. A split point is selected based on the minimization of the uncertainty of the decisions made by the classifier. Different measures of uncertainty can be used, i.e. Entropy, Gini index, and classification error [11]. All these measures have their maximum value for a uniform distribution of the objects' classification, and a minimum when all objects belong to the same class.

The random forests classifier is formed by an ensemble of decision trees which make the final decision together [6]. The main advantage of random forests classifier with respect to classic decision tree classifier is that the former one avoids the overfitting problem through the application of two pre-processing techniques: boot-

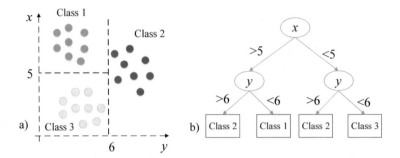

Fig. 5.2 Classification problem with three classes **a** data distribution and **b** decision tree classifier

strap aggregation, also known as bagging, and random feature selection. The former consists of generating data sets by randomly sampling with replacement from the original data set such that each decision tree learns the classification problem considering a different set of objects. The latter consists of randomly selecting a subset of features for each node of the tree without pruning. These two strategies allow to form a diverse set of trees which learn from different objects and variables of the classification problem. Therefore, a new object is classified by applying a majority voting strategy based on the outputs of all trees.

The main hyper-parameters to be selected for random forests are the uncertainty measure used to define the split points, the number of decision trees to calibrate, the number of variables randomly selected for each sub-set of the data set, and the leaf size. The latter hyper-parameter control how deep each independent tree grows, and it should be calibrated to avoid over-fitting. For instance, setting a small leaf size increases the chance of over-fitting for large training sample sizes. One significant advantage of this classifier is that in most cases the resulting decisions made during the classification process can be interpreted and analyzed. Moreover, the use of the ensemble of trees allow to create non-linear decision surfaces among the objects of different classes. The Matlab© function for using this classifier is shown in Appendix F.2.

5.3.3   Artificial Neural Networks

Artificial Neural Networks (ANNs) are classifiers inspired by the way that the brain and the human nervous system work. ANNs have become very popular dealing with a wide range of tasks such as speech recognition, computer vision, natural language processing, and others [14]. The biological mechanism is simulated through a basic building block: the artificial neuron. Mathematically, the function of a neuron is represented by

$$f(\mathbf{x}) = \phi(w_0 b + w_1 x_1 + w_2 x_2 + \cdots + w_n x_N) \tag{5.4}$$

where $\mathbf{x} = [x_1, .., x_N] \in \mathfrak{R}^N$ is the input feature vector, w_j represents the weight corresponding to input x_j (with $j = \{1, 2, \ldots, N\}$), ϕ is called activation function, and b is a bias variable with its corresponding weight w_0. This model is called the perceptron unit defined by McCulloch-Pitts [16].

A neural network is formed by different interconnected layers where each layer has a certain number of neurons that share the same inputs [7]. Usually, three types of layers are defined according to their position: input, which receives the feature information; output, where the results of the processing are given; and hidden, which are the layers between the input and output layers. A neural network with L layers is defined as a composition of L functions $f_i : E_i \times H_i \rightarrow E_{i+1}$, where E_i, H_i and E_{i+1} are inner product spaces for all $i \in L$. The vector $\mathbf{x}_i \in E_i$ represents the input to layer L_i and the vector $\mathbf{w}_i \in H_i$ represents the weights or *parameters* of layer L_i [7]. The output of the network is then calculated by

$$F(\mathbf{X}; W) = (f_L \circ \cdots \circ f_1)(\mathbf{X}) \tag{5.5}$$

where $\mathbf{X} = \{\mathbf{x}_1, \ldots, \mathbf{x}_L\}$, $W = \{\mathbf{w}_1, \ldots, \mathbf{w}_L\}$ and $\circ$ represents the convolution mathematical operation. The resulting function is a composition of the functions of the different layers.

The hyper-parameters of an ANN is the number of layers, the number of neurons in each layer, the activation function of each layer, and the evaluation/cost function. Once these hyper-parameters are set, the parameters of each neuron are learned with continuous optimization methods. The mechanism to optimize the parameters is known as back-propagation and it works as follows: the gradient of the cost function with respect to each weight is computed by the chain rule one layer at a time, and iterating backward starting from the last layer with the error between the calculated value and the pattern target value. There are several possible gradient-based optimization methods. Stochastic Gradient Descent [18] is among the most popular used to implement the backpropagation mechanism.

The main advantage of ANNs as a classifier is that it allows to model non-linear decision surfaces among the different classes thus providing an effective mechanism for pattern classification [11]. The Matlab© function for using this classifier is shown in Appendix F.3.

5.3.4 Support Vector Machines

The support vector machines (SVMs) classifier solves an optimization problem to find the optimal separating hyperplane that maximizes the margin w among the closest objects of two different classes. The objects that allow to define the hyperplane are called support vectors [19]. The resulting separating hyperplane can be represented by the following decision function $g(\mathbf{x})$

$$g(\mathbf{x}) = \mathbf{w}^T \mathbf{x} + b \tag{5.6}$$

where $\mathbf{w}$ and b (bias) are used to define a hyperplane position; and $\mathbf{x} \in \mathfrak{R}^p$ denotes a feature vector. The values of w are not restricted to form axes-parallel hyperplanes such as in decision trees. Instead, w and b are found by solving the following dual optimization problem

$$\max \; W(\mathbf{a}) = \left(\sum_{i=1}^{m} a_i - \frac{1}{2} \sum_{i,j=0}^{m} a_i a_j g_i g_j \, K(\mathbf{r_i}, \mathbf{r_j}) \right)$$

$$\text{subject to} \; \sum_i g_i a_i = 0; \quad 0 \le a_i \le C \tag{5.7}$$

where C represents the error penalty, $\mathbf{a} \in \mathfrak{R}^m$ are the Lagrange multipliers, m is the number of training examples that form the data set $X \in \{\mathbf{r_i}, \mathbf{y_i}\}^m$ with the label vector $\mathbf{y} \in \{1, -1\}$, and $K(\mathbf{r_i}, \mathbf{r_j})$ is a kernel function that allows access to spaces of higher dimensions. This is a quadratic optimization problem that is solved by the sequential minimal optimization algorithm.

Among the most popular kernel functions, due to its successful results, is the radial basis function [4, 25]

$$K(\mathbf{r_i}, \mathbf{r_j}) = \exp\left(-\gamma \|\mathbf{r_i} - \mathbf{r_j}\|^2\right) \tag{5.8}$$

Here, the term γ is usually called smoothing hyper-parameter and it defines the geometric separation of the mapped data in the high dimensional space.

SVMs are traditionally presented for solving binary classification problems and are extended to multi-class classification problems by applying discriminant strategies. The most used strategies are one-against-one and one-against-all. The hyper-parameters of a binary SVM classifier are the error penalty (C) and the smoothing hyper-parameter (γ). The main advantage of SVMs as a classifier is that it allows to create non-linear decision surfaces among the objects of different classes. The LIB-SVM library [8] is employed to train the classifier as it is shown in Appendix F.4.

5.4   Cases of Study

In this section, the pattern classification algorithms are applied to the three cases of study considered in this book. It will be shown how to calibrate the parameters of the different methods through the hyper-parameter setting strategy discussed before.

5.4.1   Continuous Stirred Tank Heater

The faults analyzed for this process are associated with sensors. Bias faults are considered in the level and temperature sensors. The goal of the classifier is then to distinguish in which sensor the fault has occurred. The assumption is that while these faults are present, measurement data of the system is collected. That data is used to train a classification tool that allows to distinguish between these two faults in the future. The three main challenges associated with data-driven classification problem are:

- The classification of faults with different magnitudes. This problem is challenging because the features of the data set used for training the classifier may depend on the magnitude of the fault. Consider the example shown in Fig. 5.3. It is illustrated that the data corresponding to the same fault may be different depending on its magnitude. However, for this example, the data of the faults is sufficiently separated such that if the fault is detected, it is possible to achieve a satisfactory classification performance regardless of the magnitude.
- The classification of faults when they occur in different operating modes. Similar to the previous challenge, the features of the data set produced by the same fault may be different from mode to mode. This situation is illustrated in Fig. 5.4. The data corresponding to a level sensor fault with the same magnitude varies depending on the operating mode. In this case, the set-point of the level in the tank varies from mode 1 (12 mA) to mode 5 (13 mA). This causes that the data of a fault in the sensor level is different for these two modes.
- The identification of whether the fault that has been detected is a new fault. This is a relevant problem for most fault diagnosis systems. There are different approaches to solve it, but this problem is out of the scope of this book. The reader can refer to the following reference for a possible solution [17].

The process of configuring a classifier is achieved by adjusting its hyper-parameters and parameters to the fault data set available. A data set of faults is assumed to be available in the first place. In order to select the best classifier, a set of candidates is initially selected. Each classifier is trained to achieve the best possible accuracy, and their performance is compared by using the Friedman and Wilcoxon non-parametric hypothesis tests.

The faults to be classified in the CSTH process are two: level and temperature sensor bias. The challenge consists of classifying the same fault under different operating modes, and different fault magnitudes, even when only fault data from one operating mode and one magnitude is available. It is assumed that initially a data set of the sensor faults with magnitude of 0.5 mA and occurring under operating mode 1 is only available. The Bayes classifier is selected to accomplish the fault classification task. If a good performance is not achieved, a more complex classifier may be required.

5.4.1.1 Results

The training accuracy with the Bayes classifier achieved 99.5%. The classifier is tested for a different data set but considering the same fault with the same magnitude and under the same operating mode. The testing accuracy achieved is 99%. The

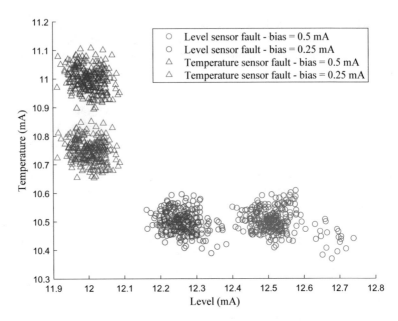

Fig. 5.3 Scatter plot of two variables of the CSTH process for different sensor fault magnitudes

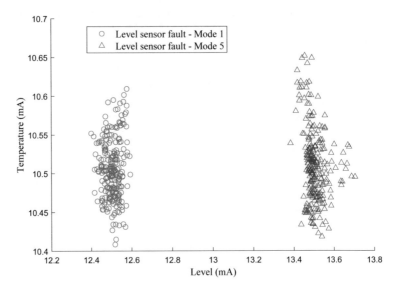

Fig. 5.4 Scatter plot of two variables of the CSTH process for a sensor fault occurring in different operating modes

confusion matrix is shown in Fig. 5.5. Only six observations are confused by the classifier. Although the testing performance is slightly lower, it is still very satisfactory. These results are obtained because the data distribution is similar for training and testing data sets, and the classes are clearly separated in the feature space. This is shown in Fig. 5.6. The classes are only mixed in the first stage of the transition from normal mode to faulty mode.

The performance results presented above are satisfactory, but only if the faults occur for operating mode 1. If the same classifier is used when a fault occurs in a different mode then its performance decreases significantly. A possible solution to build a classifier that may work for any operating mode is to scale the data with respect to the respective operating mode. This can be formalized mathematically for two data scaling approaches as follows

- Min–Max:

$$x^* = \frac{x - \min_i(x)}{\max_i(x) - \min_i(x)} \tag{5.9}$$

where $\min_i(x)$ and $\max_i(x)$ are the minimum and maximum values that characterize the variable's probability distribution during operating mode i.

- Z-score:

$$x^* = \frac{x - \mu_i}{\sigma_i} \tag{5.10}$$

Fig. 5.5 Level is different for these two modes confusion matrix for the testing data set of the CSTH process (sensor faults with same magnitude occurring in the same operating modes

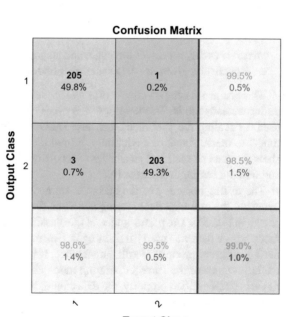

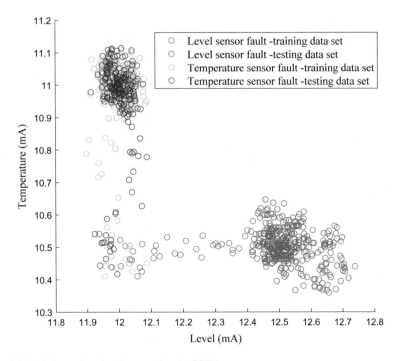

Fig. 5.6 Training and testing data sets for the CSTH process

where μ_i and σ_i represent the mean and standard deviation estimated for variable x, respectively, given the data set corresponding to operating mode i.

This pre-processing solution works well under the assumption that the fault occurs during a steady mode. This assumption is necessary because the operating mode data used for scaling the variables must have statistical features that don't change with time, i.e. mean, standard deviation, minimum, and maximum. Figures 5.7 and 5.8 show the data of the level sensor fault occurring during two operating modes with and without scaling, respectively.

The testing accuracy of the classifier trained with data of the operating mode 1 but using the min-max data scaling approach for the data of each operating mode is shown in Fig. 5.9. The performance of the classifier is the best for the same operating mode whose data was used for training it. However, the results when the faults occur during different modes are still satisfactory. It is very important to highlight again the importance of the correct operating mode identification because if the operating mode is not identified properly, the data cannot be scaled to achieve a good accuracy in the fault classification task.

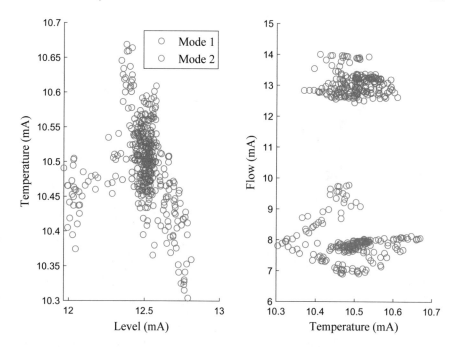

Fig. 5.7 Scatter plots of the data corresponding to the level sensor fault occurring in modes 1 and 2 without operating mode scaling for the CSTH study case

5.4.2 Hanoi Water Distribution Network

The goal of the fault classification task in the HWDN is to determine where the leak is located within the network, once it has been detected. As explained in the previous chapter, the target location usually considered for leaks are the nodes in the network.

Leakages are located by analyzing the difference between measurements and synthetic data generated through a model. There are three main approaches to leak localization depending on the type of model used: analytic, data-driven, and mixed-model. Leak localization approaches that use analytic models of the network, must first calibrate the parameters of a physics-based model that describes the main laws that characterize the behavior of such system. The main difficulty in using this approach is that developing an accurate model is labor-intensive and time-consuming. Moreover, the parameters of the networks change with time such that these models must be re-calibrated.

Data-driven models capture the network behavior from a representative set of measurements obtained under varying operating conditions. The main benefit of this approach is that models can be constantly updated as more data is gathered. The main disadvantage is that measurements are generally limited to a few sensors installed in the network. Mixed-model approaches combine the best features of the

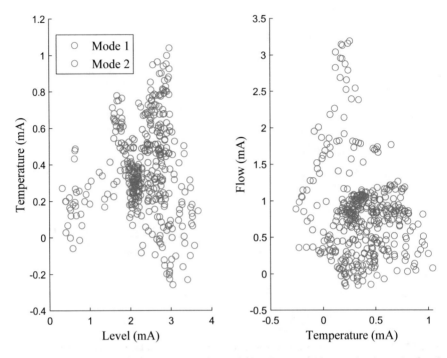

Fig. 5.8 Scatter plots of the data corresponding to the level sensor fault occurring in modes 1 and 2 with operating mode scaling for the CSTH study case

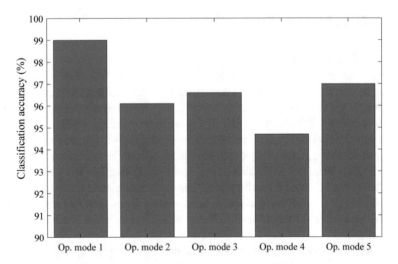

Fig. 5.9 Testing accuracy of the classifier trained with data of the operating mode 1

above-mentioned approaches. The data-driven approach models the uncertainties in the data that cannot be expressed by the physics-based model.

After a leak is detected, the fault classification mechanism is triggered. A classification tool has to be configured in a similar way as the operating mode identification and fault detection methods. This means that a classifier has to be adjusted for each time of the day. The challenge for the fault classification of this type of process is that there are many sources of uncertainty, the number of classes is high (as large as the number of nodes, which is 31 for the HWDN), and the number of measured variables is small. As a consequence, even when some leaks are detected, an accurate diagnosis may not be possible. To illustrate this issue, a classification tool will be trained by using data of leaks that have been detected (emitter coefficient around 30).

5.4.2.1 Results

The four classifiers studied in this chapter are tested. The hyper-parameters of each classifier (except for the Bayes classifier) are tuned through a grid-search according to the following parameter space

- Random Forests: (a) number of decision trees: and (b) leaf size: . Given that only three variables are measured the number of variables randomly selected for each sub-set of the data is two. The uncertainty measure used as the split criterion is Gini's diversity index.
- Artificial Neural Networks: One hidden layer and the number of neurons in the hidden layer is set to twice the number of classes. The number of neurons is increased until no further performance improvement is achieved.
- Support Vector Machines: (a) error penalty $C \in 2^\eta$, $\eta \in [-2, 5]$ and (b) smoothing parameter $\gamma \in 2^\eta$, $\eta \in [-5, 3]$.

A box-plot of the cross-validation accuracy (with 10 partitions) is shown in Fig. 5.10 for the four classifiers. The best hyper-parameters obtained for each classifier are: (a) Random Forests: 100 trees with a leaf size of 5, (b) ANN: 60 neurons in the hidden layer, (c) SVM: $C = 16$ and $\gamma = 4$. None of the four classifiers achieves a satisfactory performance. The main reason behind this poor performance is the number of uncertainties that make the classification process more difficult.

There are different strategies that may be used to partially reduce the effect of uncertainties in the classification process. Two practical solutions are:

- Filtering the measurements data as a pre-processing step. Filtering techniques are usually applied in the field of signal processing to remove undesired features from a signal [20]. Filtering of a signal can be applied either in the time or in the frequency domain. In the case of WDN data, simple filtering techniques are good enough to remove some of the unwanted uncertainty. Specifically, mean filters can be applied. A mean filter is applied in the time domain and its purpose is to replace each observation by the mean of a number of neighboring observations. The neighbors are considered with respect to the time when the observation is

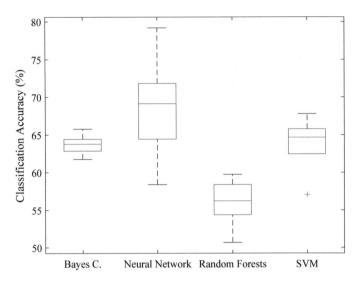

Fig. 5.10 Cross-validation classification accuracy for the Hanoi WDN by using raw data

acquired. The main benefit of filtering the data through a mean filter is that the resulting distribution approximates a Gaussian distribution, given the central limit theorem: if a random variable is the outcome of a summation of a number of independent random variables, its probability distribution function approaches the Gaussian function as the number of summands tends to infinity [23].

- Bayesian analysis of the sequential predictions made by the classifiers [22]. The prediction of a classifier can be described in terms of the probability of a class given the observation. This probability can be expressed in terms of the Bayes rule for a class ω_z as follows

$$P(\omega_z|\mathbf{x}) = \frac{P(\mathbf{x}|\omega_z)P(\omega_z)}{P(\mathbf{x})} \tag{5.11}$$

where $P(\omega_z|\mathbf{x})$ is called posterior probability that the fault ω_z is caused by the observation $\mathbf{x}$, $P(\mathbf{x}|\omega_z)$ is the probability of observation $\mathbf{x}$ if fault ω_z occurs, $P(\omega_z)$ is the prior probability for fault ω_z, and $P(\mathbf{x})$ is a normalizing factor given by the total probability law

$$P(\mathbf{x}) = \sum_{i=1}^{Z} P(\mathbf{x}|\omega_z)P(\omega_z) \tag{5.12}$$

At the beginning of the fault classification process there is no information about the prior probability of any of the leaks in the WDN. Thus, all leak locations are equiprobable. However, as the classifier makes predictions in the form of probable classes, the information from previous observations can be used to compute the current probability. As more observations are gathered, a more accurate decision

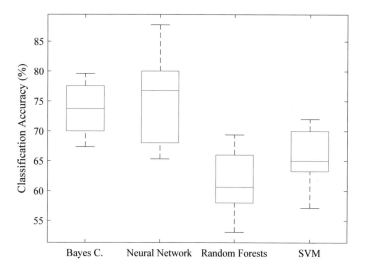

Fig. 5.11 Cross-validation classification accuracy for the Hanoi WDN by using filtered data

can be made. Therefore, a set of observations can be used to decide the node where the leak is occurring. The Matlab© implementation of the Bayesian analysis of a sequence of prediction probabilities is shown in Appendix F.5.

A box-plot of the cross-validation accuracy (with 10 partitions) is shown in Fig. 5.10 for the four classifiers, after the data has been filtered with a mean filter of 3 observations. A direct comparison of the results shown in Figs. 5.10 and 5.11 indicates an improvement in the classification accuracy for all classifiers. Random Forests has the lowest performance among all classifiers. The main reason is that this classifier achieves its best performance compared to other classifiers when the dimension of the feature space is large. These results demonstrate empirically the benefit of filtering the data as a pre-processing step for leak localization tasks (Fig. 5.11).

Figure 5.12 shows the testing classification accuracy for the four classifiers when using filtered data, and the Bayesian analysis of the classifier's output probabilities. The performance improves as more observations are analyzed to make a decision regarding the leak location. Even with Random Forests having a low performance at first, all classifiers achieve a similar performance after 16 observations are analyzed. This result highlights the importance of the Bayesian post-processing of the classifier's outputs. The Bayesian processing comes at the cost of having to wait more time to make a decision. Waiting for 16 observations means the plant operator has to wait for 8 hours for a decision regarding the leak location in the network. Nonetheless, a good compromise can be established to achieve a satisfactory performance in the less possible time period.

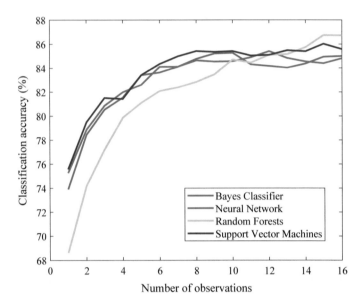

Fig. 5.12 Testing classification accuracy for the Hanoi WDN by using filtered data and Bayesian analysis of the classification results

5.4.3   Tennessee Eastman Process

The TEP brings challenges to the fault classification task that none of the previous systems had: the dimension of the feature space is large and the fault types is diverse (from step to incipient faults). Moreover, the TEP is a highly non-linear system. These characteristics make the classification of faults a complex task for this process. The fault types that will be considered are step, random variation, incipient, sticking of valves, and unknown as described in Chap. 2. Moreover, the classification task is challenging because faults can occur in different operating modes.

Two data sets of the six fault cases analyzed in this book are generated both during mode 1 and mode 3 (one for training, and one for testing). The fault classification during transitions is a challenging task that is out of the scope of this book. The reader can refer to the following reference for more information on this subject [1].

5.4.3.1   Results

The four classifiers studied in this chapter are tested. The hyper-parameters of each classifier (except for the Bayes classifier) are tuned through a grid-search according to the following parameter space

- Random Forests: (a) number of decision trees: and (b) leaf size: . The number of variables randomly selected for each sub-set of the data is selected as the square

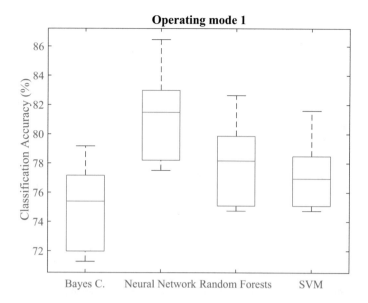

Fig. 5.13 Cross-validation classification accuracy for the TEP for operating mode 1

root of all variables of the system. The uncertainty measure used as the split criterion is Gini's diversity index.

• Artificial Neural Networks: One hidden layer and the number of neurons in the hidden layer is set initially to twice the number of classes. The number of neurons is increased until no further performance improvement is achieved.

• Support Vector Machines: (a) error penalty $C \in 2^{\eta}$, $\eta \in [-2, 5]$ and (b) smoothing parameter $\gamma \in 2^{\eta}$, $\eta \in [-5, 3]$. The LIBSVM library was employed for this purpose [8].

It must be remarked that the performance for each possible parameter configuration is evaluated based on the cross-validation error, which is calculated based on the training data set. A box-plot of the cross-validation accuracy (with 10 partitions) is shown in Figs. 5.13 and 5.14 for the four classifiers considering the faults occurring in modes 1 and 3, respectively. The cross-validation classification accuracy of all classifiers is satisfactory for operating mode 3 but the results are not good for operating mode 1.

The performance results are worse when the testing data set is used for evaluation as shown in Table 5.1. The best classification accuracy for the testing data set is achieved with the Bayes classifier and Random Forests for operating modes 1 and 3, respectively. This fact indicates that the performance of a given classifier may depend on the operating mode of the process, highlighting again the importance of the correct mode identification.

The two techniques used to improve the classification accuracy for the Hanoi WDN case study are filtering and Bayesian processing of the results. The filtering technique does not have a significant impact in the performance of the classifiers for

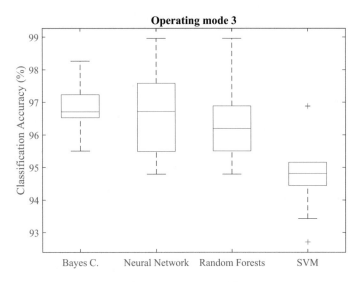

Fig. 5.14 Cross-validation classification accuracy for the TEP for operating mode 3

this case study because the effect of uncertainty is not as significant as in the Hanoi WDN. The Bayesian processing applied over a window of 15 min of data improves the performance as it shown in Table 5.2. The performance of all classifiers improves significantly in most cases. If a larger set of observations is analyzed these results may be even better, but at the expense of waiting more time for a diagnosis.

Table 5.1 Classification accuracy (%) for the testing data set of the TEP in different modes

Classifier	Operating mode 1	Operating mode 3
Bayes classifier	77.5	79.4
Neural network	69.8	73.7
Random forests	52.3	84.8
Support vector machines	63.7	71.2

Table 5.2 Classification accuracy (%) for the testing data set of the TEP in different modes with Bayesian processing

Classifier	Operating mode 1	Operating mode 3
Bayes classifier	82.45	82.21
Neural network	80.21	78.26
Random forests	53.9	89.58
Support vector machines	72.93	74.98

Finally, it must be remarked that the performance of the classifiers is not the same for both operating modes. In these experiments, each classifier is trained and evaluated with data from the same operating mode. However, if the classifiers are trained with data from one operating mode and evaluated on the other, the results are not satisfactory. The scaling technique that was applied to the CSTH process do not work well in this case because of the complexity of the system, and the nature of the faults that occur. Further research is needed to develop data-driven classifiers that can be trained with data from one operating mode and still have a good performance during the occurrence of other modes.

Remarks

Fault classification with data-driven methods is a difficult task to be accomplished for industrial systems. Fault classification is usually tackled as a pattern recognition problem where each fault represents a pattern in the data set. The task of fault classification consists then in the design of data-driven tools that recognize which fault is occurring in the system once it has been detected. The hyper-paramters and parameters of each data-driven classification method determine its performance. It is therefore of crucial importance to appropriately adjust their values. Moreover, there are data processing techniques that can be applied to improve the classification performance. Filtering and Bayesian processing are examples of pre-processing and post-processing techniques that can be applied for this purpose.

The main complexity of using data-driven classification methods for system with multiple operating modes is that the classifiers trained for one operating mode may not perform well when the same fault occurs in a different operating mode. This implies that data from the same fault across the different operating modes may be required to design a classifier for each mode. When the system is not complex and the faults analyzed are sensor faults, pre-processing techniques can be used to overcome this challenge. This was illustrated for the CSTH process. However, for more complex systems like the Tennessee Eastman Process, this approach does not work well.

References

1. Acevedo-Galán, D.L., Quiñones-Grueiro, M., Prieto-Moreno, A., Llanes-Santiago, O.: A new approach for fault diagnosis of industrial processes during transitions. In: Hernández Heredia, Y., Milián Núñez, V., Ruiz Shulcloper, J. (eds.) Progress in Artificial Intelligence and Pattern Recognition, pp. 342–350. Springer, Cham (2018)
2. Bergstra, J., Bengio, Y.: Random search for hyper-parameter optimization. J. Mach. Learn. Res. **13**(10), 281–305 (2012). http://jmlr.org/papers/v13/bergstra12a.html
3. Bergstra, J.S., Bardenet, R., Bengio, Y., Kégl, B.: Algorithms for hyper-parameter optimization. In: Shawe-Taylor, J., Zemel, R.S., Bartlett, P.L., Pereira, F., Weinberger, K.Q. (eds.) Advances in Neural Information Processing Systems, vol 24, pp 2546–2554. Curran Associates (2011)
4. Bernal-de Lázaro, J., Prieto-Moreno, A., Llanes-Santiago, O., Silva-Neto, A.: Optimizing Kernel methods to reduce dimensionality in fault diagnosis of industrial systems. Comput. Ind. Eng. **87**, 140–149 (2015)
5. Bishop, C.: Pattern Recognition and Machine Learning. Springer (2006)

6. Breiman, L.: Random Forests. Mach. Learn. **45**(1), 5–32 (2001)
7. Caterini, A.L., Chang, D.E.: Deep Neural Networks in a Mathematical Framework. Springer (2018). https://doi.org/10.1007/978-3-319-75304-1
8. Chang, C.C., Lin, C.J.: LIBSVM : a library for support vector machines. ACM Trans. Intell. Syst. Technol. **2**(3), 27:1–27:27 (2011)
9. Chawla, N.V.: Data Mining for Imbalanced Datasets: An Overview, pp 875–886. Springer, Boston, MA (2010)
10. Devijver, P.A., Kittler, J.: Pattern Recognition: A Statistical Approach. Prentice Hall, London (1982)
11. Dougherty, G.: Pattern Recognition and Classification. Springer, New York (2013)
12. Heijden, F.V.D., Duin, R., Ridder, D.D., Tax, D.: Classification, Parameter Estimation and State Estimation. Wiley (2004)
13. Li, A., Spyra, O., Perel, S., Dalibard, V., Jaderberg, M., Gu, C., Budden, D., Harley, T., Gupta, P.: A generalized framework for population based training. CoRR (2019). arxiv:abs/1902.01894
14. Liu, W., Wang, Z., Liu, X., Zeng, N., Liu, Y., Alsaadi, F.: A survey of deep neural network architectures and their applications. Neurocomputing **234**(19) (2017). https://doi.org/10.1016/j.neucom.2016.12.038
15. Miikkulainen, R., Liang, J.Z., Meyerson, E., Rawal, A., Fink, D., Francon, O., Raju, B., Shahrzad, H., Navruzyan, A., Duffy, N., Hodjat, B.: Evolving deep neural networks. CoRR (2017). arxiv:abs/1703.00548
16. Patan, K.: Artificial Neural Networks for the Modelling and Fault Diagnosis of Technical Processes. Springer, Berlin (2008)
17. Rodríguez-Ramos, A., Silva Neto, A.J., Llanes-Santiago, O.: An approach to fault diagnosis with online detection of novel faults using fuzzy clustering tools. Expert Syst. Appl. **113**, 200–212 (2018). https://doi.org/10.1016/j.eswa.2018.06.055. http://www.sciencedirect.com/science/article/pii/S0957417418304135
18. Ruder, S.: An overview of gradient descent optimization algorithms. CoRR (2016). arxiv:abs/1609.04747
19. Scholkopf, B., Smola, A.: Learning with Kernels: Support Vector Machines, Regularization, Optimization, and Beyond. MIT Press (2002)
20. Shenoi, B.A.: Introduction to Digital Signal Processing and Filter Design. Wiley (2005)
21. Snoek, J., Larochelle, H., Adams, R.P.: Practical Bayesian optimization of machine learning algorithms. In: Pereira, F., Burges, C.J.C., Bottou, L., Weinberger, K.Q. (eds.) Advances in Neural Information Processing Systems, vol 25, pp 2951–2959. Curran Associates (2012)
22. Soldevila, A., Fernandez-canti, R.M., Blesa, J., Tornil-sin, S., Puig, V.: Leak localization in water distribution networks using Bayesian classifiers. J. Process Control **55**, 1–9 (2017)
23. Theodoridis, S., Koutroumbas, K.: Pattern Recognition. Elsevier (2009)
24. Venkatasubramanian, V., Rengaswamy, R., Kavuri, S.N., Yin, K.: A review of process fault detection and diagnosis part III: process history based methods. Comput. Chem. Eng. **27**, 327–346 (2003)
25. Zhang, Q., Wu, Z.Y., Zhao, M., Qi, J., Huang, Y., Zhao, H.: Leakage zone identification in large-scale water distribution systems using multiclass support vector Machines. J. Water Resour. Plan. Manage. **142**(11), 1–15 (2016)

Chapter 6
Final Remarks

The data-driven based fault diagnosis approach for multimode continuous processes is presented in this book. Multimode processes are defined from a data-driven perspective and a multimode process assessment algorithm to determine if a process is multimode is presented. The algorithm is tested in three different benchmark problems which are extensively used throughout the book: a continuous stirred tank heater, the Hanoi water distribution network, and the Tennessee Eastman Process.

A methodology is proposed to design a data-driven fault diagnosis/monitoring loop in the general case. This methodology is also adapted to diagnose faults in multimode processes which require an operating mode identification step. Therefore, the data set of a multimode process must be processed to identify the number of operating modes as well as its type. There are different clustering methods which can be used to accomplish this step, as described in Fig. 3.2. However, their application is not straightforward, as was demonstrated in Chap. 3 because standard clustering methods like K-means do not identify the number of operating modes correctly if the process has transitions of significant duration. In such case, window-based clustering methods usually are the best method for operating mode identification.

Once the number of operating modes and type as well as the respective observations are identified, the operating mode identification and fault detection task can be accomplished. A methodology to design multimode fault detection strategies is presented in this book. There are three possible strategies shown in Fig. 4.2: (a) to design data transformation and adaptive schemes, (b) to develop a mode selection function with multiple single-mode fault detection models, and (c) use the Bayesian fusion of multiple fault detection models. The three strategies are applied to the three benchmark problems presented in the book. The main conclusion is that if the multimode process has transitions with significant duration, the second strategy is the only one that allows to obtain satisfactory performance results.

© Springer Nature Switzerland AG 2021

M. Quiñones-Grueiro et al., *Monitoring Multimode Continuous Processes*,
Studies in Systems, Decision and Control 309,
https://doi.org/10.1007/978-3-030-54738-7_6

Data-driven methods can be used for fault cause identification if a data set characterizing the behavior of the process during the faults is available. The fault cause identification task is then framed as a pattern recognition problem where classification methods can be used to solve it. Four classification methods are presented (Bayes Classifier, Random Forest, Artificial Neural Networks, and Support Vector Machines) and applied to the three benchmark problems already mentioned. The main complexity of using such methods for fault cause identification is that faults with different magnitudes, or occurring in different operating modes might have a different pattern. Moreover, these methods cannot be used to classify fault occurring during transitions because the data distribution corresponding to the fault will be different depending on when it appears.

Overall, we would like to highlight the importance of correct operating mode identification as a previous step to achieve satisfactory fault diagnosis performance.

6.1 Future Trends

The main challenge of successfully solving the fault diagnosis problem for multimode processes with data-driven methods is that faults might present different patterns depending on the operating mode. For that reason, we consider that future works with the aim of improving the performance of the fault cause identification task in multimode processes should intend to develop diagnosis methods agnostic to the operating mode where the fault occurs. Specifically, pattern recognition tools with pre-processing techniques that would allow to classify the faults independently of the operating mode. In addition, the fault cause identification during transitions remains an open problem that deserves more attention.

Appendix A
Implementation in Matlab© of the Two Algorithms for Automatic Multimode Processes Identification

A.1 Function to Determine the Bandwidth Vector h

It is assumed that the data set only considers nominal operation and it is outlier-free.

```
function [h]=Find_h(data)
% Main function: Algorithm to estimate bandwidth h
% Author: Marcos Quinones-Grueiro
% May 2020

% % Input:
% data: data set of a single variable

% % Output:
% h: bandwidth vector to be used for each observation

n=length(data);

%%%%%%%%%%%%%%%%%%%%%%%%%%%%%%%%%%%%%%%%
% Univariate rule of thumb
%%%%%%%%%%%%%%%%%%%%%%%%%%%%%%%%%%%%%%%%
q3=quantile(data,0.75);
q1=quantile(data,0.25);
gamma=q3-q1;
stdhat=std(data);
vector=[stdhat, gamma/1.34];
[val, ~]=min(vector);
h=1.06*val*n^(-1/5);
```

© Springer Nature Switzerland AG 2021
M. Quiñones-Grueiro et al., *Monitoring Multimode Continuous Processes*,
Studies in Systems, Decision and Control 309,
https://doi.org/10.1007/978-3-030-54738-7

```
%%%%%%%%%%%%%%%%%%%%%%%%%%%%%%%%%%%%%%%%
% PDF evaluation
%%%%%%%%%%%%%%%%%%%%%%%%%%%%%%%%%%%%%%%%
pdfvalue=zeros(n,1);
datatest=sort(data);
for i=1:n
    pdfvalue(i) = pdfeval(datatest(i),datatest,h);
end


%%%%%%%%%%%%%%%%%%%%%%%%%%%%%%%%%%%%%%%%
% Estimate the vector h
%%%%%%%%%%%%%%%%%%%%%%%%%%%%%%%%%%%%%%%%
teval=(1/n)*sum(log(pdfvalue));
T=exp(teval);
c=0.5;h=zeros(n,1);
for i=1:n
    h(i)=(pdfvalue(i)/T)^(-c);
end


end
```

A.2 Function to Determine the Number of Maxima

```
function [nomaxima,maxima]=Find_maxima(data,fixed_h,...
h,epsilon,maxit,precision)
% Main function: Algorithm to estimate number of maxima
of the PDF of a variable
% Author: Marcos Quinones-Grueiro
% May 2020

% % Input:
% data: data set of a single variable
% fixed_h: fixed bandwidth value
% h: bandwidth vector
% epsilon: convergence threshold
% maxit: maximum number of iterations
% precision: number of decimals to consider

% % Output:
% nomaxima: number of maxima
% maxima: maxima of the PDF of the variable
```

```
n=length(data);

count=1;
maxima=[];
%%%%%%%%%%%%%%%%%%%%%%%%%%%%%%%%%%%%%%%%%
% Estimate the local maxima of each observation
%%%%%%%%%%%%%%%%%%%%%%%%%%%%%%%%%%%%%%%%%
for i=1:n
    localmaxima = findmaximumobservation(data(i),data,...
     fixed_h,h,epsilon,maxit);
    localmaxima=...
     ceil(localmaxima*(10^(precision)))/(10^(precision));
    if length(maxima)~=0
        if ~any(maxima == localmaxima)
            maxima(count)=localmaxima;
            count=count+1;
        end
    else
        maxima(count)=localmaxima;
        count=count+1;
    end
end
nomaxima=length(maxima);

end
```

A.2.1 Function to Estimate the Local Maxima of the PDF of a Variable for an Observation

```
function [maxima] = findmaximumvariable(x,data,...
fixed_h,h,epsilon,maxit,precision)
% Main function: Algorithm to estimate the local ...
maxima of the PDF of a variable for an observation
% Author: Marcos Quinones-Grueiro
% May 2020

% % Input:
% x: observation
% data: data set of a single variable
```

```
% fixed_h: fixed bandwidth value
% h: bandwidth vector
% epsilon: convergence threshold
% maxit: maximum number of iterations
% precision: number of decimals to consider

% % Output:
% maxima: local maxima of the PDF for an observation x

n=length(data);

x_it=x;
x_it_1=0;
it=1;
while(norm(x_it-x_it_1)>epsilon && it<maxit)
    x_it_1=x_it;
    normalization_factor=0;
    for i=1:n
       normalization_factor=...
        normalization_factor+...
         gaussiankernel((x_it-data(i))/(fixed_h*h(i)));
    end
    temp=0;
    for i=1:n
       temp=temp+...
        (gaussiankernel((x_it-data(i))/...
          (fixed_h*h(i)))*data(i))/normalization_factor;
    end
    x_it=ceil(temp*(10^(precision)))/(10^(precision));
    it=it+1;
end
maxima=x_it;

end
```

A.2.2   Univariate Gaussian Kernel Function

```
function [maxima] = gaussiankernel(x,data,...
fixed_h,h,epsilon,maxit,precision)
% Main function: Gaussian kernel function
```

```
% Author: Marcos Quinones-Grueiro
% May 2020

% % Input:
% x: observation

% % Output:
% output: Gaussian kernel value

output=(1/((2*pi)^(1/2)))*exp((-1/2)*(x'*x));

end
```

Appendix B
Matlab© Script to Calculate the Clustering Evaluation Measures

```
load data

%%%%%%%%%%%%%%%%%%%%%%%%%%%%%%%%%%%%%%
% Parameters
%%%%%%%%%%%%%%%%%%%%%%%%%%%%%%%%%%%%%%
criteria = 'silhouette' or ...
'CalinskiHarabasz' or 'DaviesBouldin'
clusteringmethod='kmeans' or 'gmm'
noclusters_interval=[init:final]

evalclusters(data,clusteringmethod,...
criteria,'KList',noclusters_interval);

%%%%%%%%%%%%%%%%%%%%%%%%%%%%%%%%%%%%%%
% Visualize the evaluation performance
%%%%%%%%%%%%%%%%%%%%%%%%%%%%%%%%%%%%%%
figure;
plot(E);
xlabel('Number of clusters') ylabel('Evaluation
measure value')

%%%%%%%%%%%%%%%%%%%%%%%%%%%%%%%%%%%%%%
% For a more detailed description check in Matlab
%%%%%%%%%%%%%%%%%%%%%%%%%%%%%%%%%%%%%%
help evalclusters
```

© Springer Nature Switzerland AG 2021
M. Quiñones-Grueiro et al., *Monitoring Multimode Continuous Processes*,
Studies in Systems, Decision and Control 309,
https://doi.org/10.1007/978-3-030-54738-7

Appendix C
Clustering Evaluation Measures

C.1 Silhouette Coefficient

```
function [sil] = Silhouettecoefficient(data,clusters)

% Main function: Function to calculate ...
 the Silhouette coefficient
% Author: Marcos Quinones-Grueiro
% May 2020

% % Input:
% data: data set
% clusters: set of clusters

% % Output:
% output: Silhouette coefficient

samplesil = silhouette(dataset,clusters);
total=0;
for
i=1:length(samplesil)
    total=total+samplesil(i);
end
sil=total/length(samplesil);

end
```

© Springer Nature Switzerland AG 2021
M. Quiñones-Grueiro et al., *Monitoring Multimode Continuous Processes*,
Studies in Systems, Decision and Control 309,
https://doi.org/10.1007/978-3-030-54738-7

C.2   Calinski Harabasz Index

```
function [CH] = CalinskiHarabasz(data,clusters)

% Main function: Function to calculate ...
the Calinski Harabasz index
% Author: Marcos Quinones-Grueiro
% May 2020

% % Input:
% data: data set
% clusters: array of clusters

% % Output:
% output: Calinski Harabasz index

n = length(data);
%total observaciones
noclusters = length(clusters);
%clases

meano=mean(data);
%%%%%%%%%%%%%%%%%%%%%%%%%%%%%%%%%%%%%%%%%
% Numerator
%%%%%%%%%%%%%%%%%%%%%%%%%%%%%%%%%%%%%%%%%
nume=0;
for i=1: noclusters
    datai=clusters(i);
    [samplesi,~]=size(datai);
    nume=nume+samplesi*norm( mean(datai)-meano)^(2);
end
nume=nume/(noclusters-1);


%%%%%%%%%%%%%%%%%%%%%%%%%%%%%%%%%%%%%%%%%
% Denominator
%%%%%%%%%%%%%%%%%%%%%%%%%%%%%%%%%%%%%%%%%
deno=0;
for i=1: noclusters
    datai=clusters(i);
    [samplesi,~]=size(datai);
    meani=mean(datai);
    for j=1:samplesi
        deno=deno+norm(datai(j,:)- meani)^(2);
    end
```

```
end
deno=deno/(n-noclusters);

%%%%%%%%%%%%%%%%%%%%%%%%%%%%%%%%%%%%%%%%%%
% Output
%%%%%%%%%%%%%%%%%%%%%%%%%%%%%%%%%%%%%%%%%%
CH=nume/deno;

end
```

C.3   Davies Bouldin Index

```
function [DB] = DaviesBouldin(data,clusters)

% Main function: Function to calculate the ...
Davies Bouldin index
% Author: Marcos Quinones-Grueiro
% May 2020

% % Input:
% data: data set
% clusters: array of clusters

% % Output:
% output: Davies Bouldin index

n = length(data); %total observaciones
noclusters = length(clusters);   %clases

DB=0;
for i=1:noclusters
    evaluation=[];
    for j=1:noclusters
        if j~=i
            datai=clusters(i);
            [samplesi,~]=size(datai);
            meani=mean(datai);
```

```
%% sumatoria 1
sum1=0;
for k=1:samplesi
    sum1=sum1+norm(datai(k,:)- meani);
end
sum1=sum1/samplesi;

%% sumatoria 2
dataj=clusters(j);
[samplesj,~]=size(dataj);
meanj=mean(dataj);
sum2=0;
for k=1:samplesj
    sum2=sum2+norm(dataj(k,:)- meanj);
end
sum2=sum2/samplesj;

%% evaluar
evaluation(j)=(sum1+sum2)/norm(meani- meanj);
        end
    end
    DB=DB+max(evaluation);
end
DB=DB/noclusters;

end
```

Appendix D
Matlab© Implementations of the Clustering Methods

D.1 K-Means

```
%%%%%%%%%%%%%%%%%%%%%%%%%%%%%%%%%%%%%%%%
% Function already implemented in Matlab
%%%%%%%%%%%%%%%%%%%%%%%%%%%%%%%%%%%%%%%%
help kmeans
```

D.2 Mixture Modeling Clustering

```
%%%%%%%%%%%%%%%%%%%%%%%%%%%%%%%%%%%%%%%%%%
% Function already implemented in Matlab
%%%%%%%%%%%%%%%%%%%%%%%%%%%%%%%%%%%%%%%%%%
help gmm
```

D.3 Window-Based Clustering

```
function [steadyclusters,transitions,idxs] = ...
WindowsClustering(data,windowlength,... assthreshold,minsteady)

% Main function: Function to perform ...
Window-based clustering where the criteria ...
to distinguish steady modes is the duration
% Author: Marcos Quinones-Grueiro
```

© Springer Nature Switzerland AG 2021
M. Quiñones-Grueiro et al., *Monitoring Multimode Continuous Processes*,
Studies in Systems, Decision and Control 309,
https://doi.org/10.1007/978-3-030-54738-7

```
% May 2020

% % Input:
% data: data set
% windowlength: length of the moving window
% assthreshold: assignment threshold
% minsteady: minimum duration to consider a steady mode

% % Output:
% steadyclusters: array of steady clusters
% transitions: array of transition clusters
% idxs: vector with the correspondence of each ...
observation with its cluster

%%%%%%%%%%%%%%%%%%%%%%%%%%%%%%%%%%%%%%%%%
% Features of the data
%%%%%%%%%%%%%%%%%%%%%%%%%%%%%%%%%%%%%%%%%
m=size(data,1);
n=size(data,2);
labels=1:m;
%array identifying each sample and the cluster it belongs to
idx=[labels',zeros(m,1)];

%%%%%%%%%%%%%%%%%%%%%%%%%%%%%%%%%%%%%%%%%
% split data set sequentially into a
% set of windows with size windowlength
% and calculate the feature vector of each cluster
%%%%%%%%%%%%%%%%%%%%%%%%%%%%%%%%%%%%%%%%%
totalwindows=m/windowlength;
count=1;
for i=1:totalwindows
    clusters{i}=Dataset(count:count+windowlength-1,:);
    clusterslabels{i}=idx(count:count+windowlength-1,:);
    features{i}=mean(clusters{i});
    count=count+windowlength;
end

%%%%%%%%%%%%%%%%%%%%%%%%%%%%%%%%%%%%%%%%%%%
% Merge clusters according to the similarity of their features
%%%%%%%%%%%%%%%%%%%%%%%%%%%%%%%%%%%%%%%%%%%
noclust=[0 totalwindows];
while(noclust(2) ~= noclust(1))
    for i=1:totalwindows
        for l=1:totalwindows
            if (l==(i+1))
                kdistfeature = norm(features{i}-features{l});
                if(kdistfeature<=assthreshold)
                    A=[clusters{i} ; clusters{l}];
                    clusters{i}=A;
                    features{i}=mean(clusters{i});
                    clusters(l)=[];
                    features(l)=[];

                    %joining cluster labels
                    A=[clusterslabels{i} ; clusterslabels{l}];
                    clusterslabels{i}=A;
```

```
                          clusterslabels(1)=[];
                          totalwindows=totalwindows-1;
                  end
              end
          end
      end
      for i=1:totalwindows
          features{i}=mean(clusters{i});
      end
      noclust(1)=noclust(2);
      noclust(2)=totalwindows;
  end


  %%%%%%%%%%%%%%%%%%%%%%%%%%%%%%%%%%%%%%%%%%
  % Merge steady clusters according to the similarity of their features
  %%%%%%%%%%%%%%%%%%%%%%%%%%%%%%%%%%%%%%%%%%
  totalclusters=length(clusters);
  noclust=[0
  totalclusters];interrupt=0;
  while(noclust(2) ~= noclust(1))
      for i=1:totalclusters
          for l=1:totalclusters
              if (length(clusters{i})>minsteady ...
               && length(clusters{l})>minsteady && i~=l)

                  kdistfeature = norm(features{i}-features{l});

                  if(kdistfeature<=assthreshold)
                      A=[clusters{i} ; clusters{l}];
                      clusters{i}=A;
                      features{i}=mean(clusters{i});
                      clusters(l)=[];
                      features(l)=[];

                      %joining cluster labels
                      A=[clusterslabels{i} ; clusterslabels{l}];
                      clusterslabels{i}=A;
                      clusterslabels(l)=[];
                      totalclusters=totalclusters-1;
                      interrupt=1;
                      break;
                  end
              end
          end
          if interrupt==1
              interrupt=0;
              break;
          end
      end
      for i=1:totalclusters
          features{i}=mean(clusters{i});
      end
      noclust(1)=noclust(2);
      noclust(2)=totalclusters;
  end
  noclusters=length(clusters);
```

```
%%%%%%%%%%%%%%%%%%%%%%%%%%%%%%%%%%%%%%%%
% Separate steady modes from transitions
%%%%%%%%%%%%%%%%%%%%%%%%%%%%%%%%%%%%%%%%
count=1;
transitions=[];
steadyclusters=[];
positionssteady=[];
labels=[];
for i=1:noclusters
    if(length(clusters{i})>minsteady)
        labels(count)=count;
        idxs=clusterslabels{i};
        idxs(:,2)=count;
        clusterslabels{i}=idxs;
        positionssteady(count)=i;
        steadyclusters{count}=clusters{i};
        featuresteady{count}=mean(clusters{i});
        count=count+1;

    else
        transitions=[transitions;clusters{i}];
    end
end
%%%%%%%%%%%%%%%%%%%%%%%%%%%%%%%%%%%%%%%%%
% Correspondence of each observation with its cluster
%%%%%%%%%%%%%%%%%%%%%%%%%%%%%%%%%%%%%%%%%
idxs=[];count=1;
for i=1:numel(clusterslabels)
    temp=clusterslabels{i};
    for j=1:length(temp)
        idxs(count,:)=temp(j,:);
        count=count+1;
    end
end

end
```

Appendix E
Matlab© Implementation of the Monitoring Methods

E.1 Estimation of the Monitoring Parameters of the Principal Component Analysis Model

```
function [T21,Q1,mu,sigma,TM,AV,l] = PCA_param( data, ...
 info_percent, confidence,normalize)
% Main function: Function to estimate the parameters of...
% the PCA monitoring model
% Author: Marcos Quinones-Grueiro
% May 2020

% % Input:
% data: data set
% info_percent: percent of information to retain
% confidence: confidence to estimate the statistics threshold
% normalize: true to normalize, otherwise false

% % Output:
% T21: Threshold for T2 statistic
% Q1: Threshold for Q statistic
% mu: mean vector of the variables
% sigma: standard deviation vector of the variables
% TM: Transformation matrix
% AV: Matrix with auto-values
% l: number of components retained

m=size(data,1);
n=size(data,2);
```

© Springer Nature Switzerland AG 2021
M. Quiñones-Grueiro et al., *Monitoring Multimode Continuous Processes*,
Studies in Systems, Decision and Control 309,
https://doi.org/10.1007/978-3-030-54738-7

```
if(normalize == true)
    [data,mu,sigma] = zscore(data);
else
    mu = mean(data);
    sigma = std(data);
end

c = cov(data);
[TMal,AVal,explained] = pcacov(c);

I=eye(m);
%%%%%%%%%%%%%%%%%%%%%%%%%%%%%%%%%%%%%%%%%%
% Determining the number of components retained
%%%%%%%%%%%%%%%%%%%%%%%%%%%%%%%%%%%%%%%%%%
init=0;
count=0;
while(init < info_percent)
    init=init + explained(count+1);
    count=count+1;
end l=count;
%obtain matrices with reduced dimension
TM= TMal(:,1:l);
AV = AVal(1:l);

%%%%%%%%%%%%%%%%%%%%%%%%%%%%%%%%%%%%%%%%%%%
% Non-parametric approach to estimate
% the statistics thresholds
%%%%%%%%%%%%%%%%%%%%%%%%%%%%%%%%%%%%%%%%%%%
for i=1:length(data)
    sample =  data(i,:);
    % Q or SPE statistic
    r=(I - TM*(TM'))*sample';
    Q(i)=(r')*r;
    %T_2 statistic
    T2(i)=(sample)*(TM* (inv(diag(AV)))*(TM'))*(sample');
end
T21=quantile(T2,confidence);
Q1=quantile(Q,confidence);

end
```

E.2 Script for the Application of the PCA Monitoring Model to a Data Set

```
% data set
load data
% structure containing all parameters of ...
%  the PCA monitoring model

load parameters

T_21=     parameters.T_21;
Q1=     parameters.Q1;
mu=     parameters.mu;
sigma=     parameters.sigma;
TM=    parameters.TM;
AV=     parameters.AV;
comp_retain =    parameters.l;
normalize =    parameters.normalize;

if normalize == true
pos=find(sigma==0);
sigma(pos)=1;
sample =   data - mu;
sample =   sample./sigma;
end

fault=zeros(length(sample),1);
for i=1:length(sample)
    observation=sample(i,:);

    %Q or SPE
r=(I - TM*(TM'))*observation';
Q=(r')*r;

%T_2
T2=(observation)*(TM* (inv(diag(AV))))*(TM'))*(observation');

%fault detection condition
if(Q>Q1 || T2>T21)
```

```
fault(i)=1;
end
end


% In the vector fault=1 means that a fault is detected
```

E.3   Estimation of the Monitoring Parameters of the Independent Component Analysis Model

```
function [Il,Iel,Ql,mu,sigma,Q, W, Wd, Bd, We] = ...
ICA_param( data, num_comp, confidence,normalize)
% Main function: Function to estimate the parameters of...
% the ICA monitoring model
% Author: Marcos Quinones-Grueiro
% May 2020

% % Input:
% data: data set
% num_comp: number of components to retain
% confidence: confidence to estimate the statistics threshold
% normalize: true to normalize, otherwise false

% % Output:
% Il: Threshold for I statistic
% Iel: Threshold for I statistic
% Ql: Threshold for Q statistic
% mu: mean vector of the variables
% sigma: standard deviation vector of the variables
% W: Demixing matrix
% Wd: Reduced demixing matrix
% We: Excluded transformation matrix
% Bd: Transformation matrix
% Q: Whitening matrix

m=size(data,1);
n=size(data,2);

if(normalize == true)
    [data,mu,sigma] = zscore(data);
else
    mu = mean(data);
    sigma = std(data);
end
```

```matlab
normal=data';
%%%%%%%%%%%%%%%%%%%%%%%%%%%%%%%%%%%%%%%%
% Running ICA algorithm
% check fastica function at
% http://research.ics.aalto.fi/ica/fastica/code/dlcode.shtml
%%%%%%%%%%%%%%%%%%%%%%%%%%%%%%%%%%%%%%%%
[whitesig, Q, DewhiteningQ] = fastica(normal, 'only', 'white');
[icasig, A, W] = fastica(normal,'verbose','off','epsilon',0.01,...
'maxNumIterations',1000,'approach','defl','g','gauss');

%%%%%%%%%%%%%%%%%%%%%%%%%%%%%%%%%%%%%%%%
% Re-order components according to information percent
%%%%%%%%%%%%%%%%%%%%%%%%%%%%%%%%%%%%%%%%
m=size(W,1);
L2total=0;
for i=1:size(W,1)
    L2n(i) = norm(W(i,:));
    L2total = L2total + L2n(i);
end
for i=1:size(W,1)
    L2nExplained(i) = L2n(i)/L2total;
end
[L2nExplainedOrd,L2Indices] = sort(L2nExplained,'descend');

%%%%%%%%%%%%%%%%%%%%%%%%%%%%%%%%%%%%%%%%
% Obtain matrices
%%%%%%%%%%%%%%%%%%%%%%%%%%%%%%%%%%%%%%%%
numOfIC=numcomp;

for i=1:numOfIC
    Wd(i,:)=W(L2Indices(i),:);
end
Bd=(Wd*inv(Q))';

%Selecting the number of excluded part of the components
for i=(numOfIC+1):m
    We(i-numOfIC,:)=W(L2Indices(i),:);
end

normal=normal';
W=W';

%%%%%%%%%%%%%%%%%%%%%%%%%%%%%%%%%%%%%%%%
% Non-parametric approach to estimate
% the statistics thresholds
%%%%%%%%%%%%%%%%%%%%%%%%%%%%%%%%%%%%%%%%
for i=1:length(normal)
    sample =  normal(i,:);
    sd_new=Wd*sample';
    I(i)=sd_new'*sd_new;
    se_new=We*sample';
    Ie(i)=se_new'*se_new;
    sample_res=inv(Q)*Bd*sd_new;
```

```
        SPE(i)=(sample'-sample_res)'*(sample'-sample_res);
end
Ilim=quantile(I,confidence);
Ielim=quantile(Ie,confidence);
Ql=quantile(SPE,confidence);

end
```

E.4   Script for the Application of the ICA Monitoring Model to a Data Set

```
% data set
load data
% structure containing all parameters of ...
%  the ICA monitoring model

load parameters

Il=      parameters.Il;
Iel=      parameters.Iel;
Ql=      parameters.Ql;
mu=      parameters.mu;
sigma=      parameters.sigma;
W=    parameters.W;
Wd=    parameters.Wd;
We=    parameters.We;
Bd=    parameters.Bd;
Q=    parameters.Q;
comp_retain =    parameters.l;
normalize =    parameters.normalize;

if normalize == true
pos=find(sigma==0);
sigma(pos)=1;
sample =    data - mu;
sample =    sample./sigma;
end

fault=zeros(length(sample),1);
for i=1:length(sample)
```

```
    observation=sample(i,:);

    %I
sd_new=Wd*observation';
I=sd_new'*sd_new;

%Ie
se_new=We*observation';
Ie=se_new'*se_new;

%Q or SPE
sample_res=inv(Q)*Bd*sd_new;
SPE=(sample'-sample_res)'*(sample'-sample_res);

%fault detection condition
if(Q>SPE1 || I>I1 || Ie>Ie1)
fault(i)=1;
end
end

% In the vector fault=1 means that a fault is detected
```

E.5   Training a Support Vector Data Description Model

```
%%%%%%%%%%%%%%%%%%%%%%%%%%%%%%%%%%%%%%
% dd_tools (Data description tools) library
% https://www.tudelft.nl/ewi/over-de-faculteit/afdelingen/ ...
intelligent-systems/pattern-recognition-bioinformatics/ ...
pattern-recognition-laboratory/data-and-software/dd-tools/
%%%%%%%%%%%%%%%%%%%%%%%%%%%%%%%%%%%%%%

%%%%%%%%%%%%%%%%%%%%%%%%%%%%%%%%%%%%%%%
% For training check the following function
% from the dd_tools library
%%%%%%%%%%%%%%%%%%%%%%%%%%%%%%%%%%%%%%%
svdd
```

E.6   Script for the Application of the SVDD Monitoring Model to a Data Set

```
% data set
load data

% structure containing all parameters of ...
%          the SVDD monitoring model
load parameters

% Support Vectors
SV=      parameters.SV;
% Lagrange multipliers
alpha=      parameters.alpha;
% Kernel bandwidth
bandwidth=      parameters.bandwidth;
% Detection threshold - Radius of the hyper-sphere in ..
% high dimensional space
Rl=      parameters.Rl;
normalize =    parameters.normalize;

if normalize == true
pos=find(sigma==0);
sigma(pos)=1;
sample =    data - mu;
sample =    sample./sigma;
end

[Rows,Columns] = size(SV);
KM = zeros(Rows,Rows);
for i=1:Rows
    for j=1:Rows
        s = SV(i,:) - SV(j,:);
        KM(i,j) = exp(-(norm(s)^2)./(bandwidth^2));
    end
end

L = size(alpha,2);
TM = [data' SV']';

% Calculate Kernel matrix
[Rows,Columns] = size(TM);
KTM = zeros(Rows,Rows);
for i=1:Rows
```

```matlab
        for j=1:Rows
            s = TM(i,:) - TM(j,:);
            t = norm(s);
            KTM(i,j) = exp(-(t^2)/(bandwidth^2));
        end
    end
end
TM=KTM;

[tR tC] = size(data);
[sR sC] = size(SV);
alph_i = zeros(1,sR);
sub1
= 0;
ii = 0;
for i=1:L
    if (alpha(i)>0)
        ii = ii+1;
        alph_i(ii) = alpha(i);
    end

    for j=1:L
        if ((alpha(i)>0)&&(alpha(j)>0))
            sub1 = sub1 + alpha(i) * alpha(j) * KM(i,j);
        end
    end
end

out = zeros(1,tR);
for i=1:tR
    sub2 = 0;
    for j=1:sR
        sub2 = sub2 + alph_i(j) * TM(i,tR+j);
    end
    sub2 = sqrt(1 -2 * sub2 + sub1);
    R = sub2;

    %fault detection condition
if(R>R1)
fault(i)=1;
end
end

% In the vector fault=1 means that a fault is detected
```

Appendix F
Matlab© Implementation of Fault Classification Methods

F.1 Bayesian Classifier

```
%%%%%%%%%%%%%%%%%%%%%%%%%%%%%%%%%%%%%%%%
% Function already implemented in Matlab
%%%%%%%%%%%%%%%%%%%%%%%%%%%%%%%%%%%%%%%%
help fitcdiscr
```

F.2 Random Forests

```
%%%%%%%%%%%%%%%%%%%%%%%%%%%%%%%%%%%%%%%%
% Function already implemented in Matlab
%%%%%%%%%%%%%%%%%%%%%%%%%%%%%%%%%%%%%%%%
help TreeBagger
```

F.3 Artificial Neural Networks

```
%%%%%%%%%%%%%%%%%%%%%%%%%%%%%%%%%%%%%%%%
% Function already implemented in Matlab
%%%%%%%%%%%%%%%%%%%%%%%%%%%%%%%%%%%%%%%%
help patternnet
```

© Springer Nature Switzerland AG 2021   151
M. Quiñones-Grueiro et al., *Monitoring Multimode Continuous Processes*,
Studies in Systems, Decision and Control 309,
https://doi.org/10.1007/978-3-030-54738-7

F.4   Support Vector Machines

```
%%%%%%%%%%%%%%%%%%%%%%%%%%%%%%%%%%%%%%%%%
% Check the following function in LIBSVM library
% (https://www.csie.ntu.edu.tw/~cjlin/libsvm/)
%%%%%%%%%%%%%%%%%%%%%%%%%%%%%%%%%%%%%%%%%
svmtrain
```

F.5   Bayesian Post-processing Method

```
%%%%%%%%%%%%%%%%%%%%%%%%%%%%%%%%%%%%%%%%%
% Post-processing function
%%%%%%%%%%%%%%%%%%%%%%%%%%%%%%%%%%%%%%%%%
function [ probability,class ] = BayesianProcessing( probabilities )
% Main function: Function to apply Bayesian ...
processing to a set of probabilities
resulting from the predictions of a classifier
% Author: Marcos Quinones-Grueiro
% May 2020

% % Input:
% probabilities (m x n): prediction probabilities for each class
% m: number of classes
% n: number of observations

% % Output:
% class: most probable class
% probability: probability value corresponding to the most probable class

for sample_no = 1:size(probabilities,2)
    if sample_no == 1
        % probability a posteriori with equal a priori probability

        TotProb=0;
        for tot = 1:size(probabilities,1)
            TotProb = TotProb + probabilities(tot,sample_no);
        end

        for tot = 1:size(probabilities,1)
            Prob(tot,sample) = probabilities(tot,sample)/TotProb;
        end
    else
        % probability a posteriori considering a priori probability

        TotProb=0;
        for tot = 1:size(probabilities,1)
            TotProb = TotProb + probabilities(tot,sample_no) ...
                                        *Prob(tot,sample_no-1);
        end
```

```
            for tot = 1:size(probabilities,1)
                Prob(tot,sample_no) = probabilities(tot,sample_no) ...
                                    *Prob(tot,sample_no-1)/TotProb;
            end
        end
end
[probability,class] = max(Prob(:,size(probabilities,2)));
end
```

Printed in the United States
by Baker & Taylor Publisher Services